Edward Martin

Essentials of Surgery

Together with a full description of the handkerchief and roller bandage ; arranged in the form of questions and answers, prepared especially for students of medicine.

Second Edition

Edward Martin

Essentials of Surgery

Together with a full description of the handkerchief and roller bandage ; arranged in the form of questions and answers, prepared especially for students of medicine. Second Edition

ISBN/EAN: 9783337426651

Printed in Europe, USA, Canada, Australia, Japan

Cover: Foto ©berggeist007 / pixelio.de

More available books at **www.hansebooks.com**

SAUNDERS' QUESTION-COMPENDS. No. 2.

ESSENTIALS OF SURGERY.

TOGETHER WITH A

FULL DESCRIPTION OF THE HANDKERCHIEF
AND ROLLER BANDAGE.

ARRANGED IN THE FORM OF

QUESTIONS AND ANSWERS

PREPARED ESPECIALLY FOR

STUDENTS OF MEDICINE.

BY

EDWARD MARTIN, A.M., M.D.,
INSTRUCTOR OF OPERATIVE SURGERY, UNIVERSITY OF PENNSYLVANIA; SURGEON TO THE
HOWARD HOSPITAL; SURGEON TO THE OUT-PATIENT DEPARTMENTS OF
THE UNIVERSITY AND CHILDREN'S HOSPITALS.

SECOND EDITION. WITH NINETY ILLUSTRATIONS.

PHILADELPHIA:
W. B. SAUNDERS,
TENTH AND CHESTNUT STREETS.
LONDON: HENRY RENSHAW.   MELBOURNE: GEORGE ROBERTSON & CO.
1889.

J. A. Carveth & Co.,
Toronto, Ont.

PREFACE.

As one thrown yearly in contact with large numbers of medical students, and familiar with the furious rate at which they are driven, the writer feels assured that, under our present system of rapid education, outline works are of distinct value. Third year men who attend six lectures and two clinics daily have no time for reading, no time for systematizing their knowledge on any one subject. This work must either be done for them, or left undone. The author has carefully gone over the subject of Surgery, and has endeavored to emphasize the essential points as a framework upon which more detailed knowledge may be hung. Agnew, Ashhurst, Gross, Walsham, Tillmann, König, Treves, Weir, Smith, Gerster, and many others have been freely consulted. The table of Urinary Calculi is taken direct from Moullin's article in Treves's manual. The classification of Venereal Diseases follows that of White (University of Pennsylvania). To Mr. W. L. Alrich and Mr. Daniel Webster thanks are due for their valuable assistance.

The author has made an earnest effort to be accurate, concise, and modern.

E. M.

October 10, 1888.

CONTENTS.

	PAGE
Inflammation	17
Abscess	27
Ulceration	31
Mortification	38
Wounds	44
The germ theory of	44
Shock	45
Wound fever	47
Erysipelas	50
Tetanus	52
Hydrophobia	54
Glanders	55
Malignant pustule	55
The healing of wounds	56
The treatment of wounds	57
Wounds of arteries	73
Wounds of nerves	75
Head injuries	75
Injuries of the meninges and brain	81
Concussion and contusion	83
Compression	84
Intracranial inflammation	85
Cerebral localization	87
Wounds of the face	90
Wounds of the neck	91
Wounds of the chest	92
Wounds of the abdomen	95
Burns and scalds	102

CONTENTS.

	PAGE
Fractures	105
Special fractures	112
Luxations or dislocations	137
Special luxations	140
Sprains	158
Wounds of joints	159
Synovitis	160
Arthritis	161
Coxalgia	163
Sacro-iliac disease	166
White swelling of the knee-joint	166
Rheumatoid arthritis	167
Loose bodies in joints	167
Anchylosis	168
Diseases of bones	169
Periostitis	169
Osteitis	170
Osteomyelitis	170
Abscess of bone	171
Caries	172
Necrosis	172
Tubercle	173
Syphilitic bone disease	173
Osteomalacia	174
Pott's disease	174
Rickets	176
Hæmophilia	177
Struma	177
Curvature of the spine	177
Hernia	179
Special hernias	188
Intestinal obstruction	196
Diseases of the anus and rectum	198
Syphilis	206
Chancroid	210
Gonorrhœa	211

CONTENTS.

	PAGE
Urethral deformities	217
Stricture of the urethra	217
Diseases of the prostate	224
Affections of the bladder	227
Rupture of the bladder	227
Exstrophy of the bladder	227
Cystitis	228
Atony and paralysis of the bladder	229
Hæmaturia	229
Retention of urine	230
Stone in the bladder	233
Hydrocele	238
Hæmatocele	239
Varicocele	240
Sarcocele	240
Diseases of veins	242
Angioma	244
Aneurism	245
Diseases of the lymphatics	248
Effects of cold	249
Foreign body in the air-passages	250
Affections of the œsophagus	251
Surgical affections of the breast	253
Club-foot	254
Hare-lip and cleft palate	255
Diseases of bursæ and tendons	256
Bursitis	256
Onychia	257
Anæsthetics	258
Ligation of arteries	261
Excision of joints	278
Amputations	282
Bandaging	290
The roller bandage	290
Head bandage	299
Handkerchiefs	301

ESSENTIALS OF SURGERY.

INFLAMMATION.

What is inflammation?

Inflammation is a perversion of nutrition attended with *redness, heat, swelling, pain,* and a *tendency to exudation.*

Name the varieties of inflammation.

1. Acute. 2. Chronic.

Name the causes of inflammation.

1. *Predisposing.* Anything lowering the powers of resistance, such as heredity, age, sex, occupation, habits, food, previous inflammation, temperature, climate, temperament, mental condition.

2. *Exciting.* Traumatism, heat, cold, acids, alkalies, micro-organisms and their products.

How does inflammation extend?

By the means of bloodvessels or lymphatics. Extension by *continuity, contiguity, metastasis,* and *sympathy* is really due to either the blood or lymph vessels.

How may inflammation terminate?

1. Resolution, or return of tissues to their normal condition.
2. Organization, or new formation.
3. Death of tissue, by suppuration or mortification.

What are the phenomena of inflammation?

1. *Disturbed innervation,* causing, first, a contraction of the capillaries, followed shortly by a paralytic dilatation producing active hyperæmia.

2. *Alteration in the bloodvessels and contents.* The vascular walls are widely dilated, plastic, and their epithelium greatly swollen. The white blood corpuscles are numerous, cling to the sides, and the current is slowed or stopped. The red corpuscles stick together; the liquor sanguinis contains more fibrin forming elements.

3. *Exudation* or passage through the walls of white corpuscles (diapedesis) and liquor sanguinis.

4. *Alteration in the perivascular tissue.* Intercellular matrix undergoes mucoid softening, connective-tissue corpuscles proliferate, the exudate coagulates.

What zones are found about an inflamed area?

Most peripherally, a bright red ring where the bloodvessels are widened, called the zone of *determination*. Within this an area in which from overcrowding the blood current is slow, the color here is somewhat dusky, this area is called the *zone of congestion*. Centrally, an area where the blood current is practically at a stand-still, this is the focus of inflammation, and is termed the *zone of stasis*.

What are the stages of inflammation?

First stage. Acute hyperæmia with slight exudation.

Second stage. Lymphatization or free exudation and the formation of plastic lymph.

Third stage. Suppuration or formation of pus due to the death of white blood corpuscles and their fibrinous trabeculæ.

What is plastic lymph?

The exudate of acute inflammation. It is made up of white blood corpuscles and proliferated connective-tissue cells, imbedded in a frame-work of coagulated fibrin.

Name the different kinds of exudate.

1. Serous. Thin, non-organizable. Examples: hydrocele, ascites, hydrothorax.

2. Fibrinous. Contains much fibrin, coagulates, and readily undergoes organization.

How may the various stages of inflammation terminate?

Active hyperæmia may terminate in resolution or in exudation.

Exudation may terminate in resolution, organization, or suppuration.

Suppuration may terminate in granulation or in death of the part.

Describe resolution.

The dilated vessels again contract, the white blood corpuscles begin to move away from the inflamed area as circulation is restored. The migrated corpuscles either return to the bloodvessels, degenerate, and are carried off by the lymphatics, or remain as fixed connective-tissue corpuscles. The fibrin becomes granular and is absorbed.

Describe organization.

New bloodvessels are formed in the exudate by looping of the old ones; these loops anastomose with each other, forming a network. In addition new vessels are separately developed in the inflammatory tissue which, in turn, anastomose with the previously existing vessels. If the irritation ceases many of the exudation cells disintegrate and are removed, others are converted into connective-tissue corpuscles, which, by their contraction, obliterate the new bloodvessels and form *cicatrices*.

Describe suppuration.

If, from great irritation, the exudation is excessive there will be acute starvation, with disintegration of the central portions from occlusion of the supplying bloodvessels. If this dead central portion be kept aseptic it may be absorbed; if, however, any of the pyogenic organisms gain access to it they keep up the irritation, cause additional effusion, and produce *suppuration*.

What is pus?

Pus is the product of suppuration. It is a creamy looking, highly albuminous liquid, sp. gr. 1030, and contains fat, blood salts, tyrosin, leucin, and other nitrogenous derivatives. On standing it separates into *liquor puris*, a clear liquid, practically the same as liquor sanguinis, and *pus corpuscles*, made up of living or dead leucocytes.

Name the varieties of pus.

Laudable. Thick and cream-like: this variety comes from ordinary acute inflammation in healthy subjects.

Sanious. Thin, reddish, mixed with blood. From malignant disease, chronic ulcers, etc.

Ichorous. Thin, watery, irritating. From chronic ulcers, bone disease, etc.

Curdy or cheesy. Contains flakes of degenerated fibrin. From chronic abscesses connected with bone disease.

Gummy. Thick and ropy. From syphilitic abscesses.

Contagious pus. Muco pus, etc.

What becomes of pus?

It may be disintegrated and absorbed; it may be discharged; its more liquid portions may be absorbed, while the solid portions, together with the affected tissues, undergo fatty disintegration and remain as a putty-like mass, this constitutes *caseation*.

Name the varieties of suppuration.

Circumscribed. Diffuse. The diffuse may be *superficial* as in the cases of coryza and dysentery; or *deep* as in cellulitis.

What are the symptoms of acute inflammation?

Fever, together with *redness, heat, swelling, pain, alteration of function and nutrition.*

What are the characteristics of inflammatory redness?

It is *persistent;* if the capillaries are emptied by pressure with the finger the redness *instantly returns* on removal of the pressure. The shade of color depends upon the rapidity and freedom of the circulation; if dark or bluish it denotes obstruction or stasis. Copper-red often denotes syphilitic inflammation. Rose-red streaks along the course of the lymph vessels denote *lymphangitis*. A dusky-red tract in the course of a vein indicates *phlebitis*.

At what portion of an inflammatory area is heat most marked?

At the focus or centre.

Describe inflammatory swelling.

It is due to the increased amount of blood in the part, to proliferation, and to exudation. It is soft in acute, hard in chronic inflammations; is especially well marked in loose connective tissues. Its limitations by fascia may indicate the seat of inflammation.

Describe inflammatory pain.

It is *persistent, increased by pressure or motion*, and accompanied by the *signs of inflammation*. Is mainly due to mechanical injury to the *nerves* from the swelling. Most intense in dense structures. May be felt in regions remote from the inflamed area; instance, the knee pain of coxalgia or the shoulder pain of hepatitis.

Describe inflammatory alteration of function.

Secretions are perverted or abolished. Reflexes become greatly exaggerated; instance, the tenesmus (straining) of dysentery, the strangury of cystitis, the convulsions of teething. Non-sensitive parts become hyper-sensitive; instance, the pain of peritonitis or of teething.

Describe the constitutional symptoms of inflammation.

Fever. May be *sthenic* or *asthenic* in type.

1. *Sthenic inflammatory fever.*

a. Circulatory symptoms. Full, strong, rapid pulse, flushed face, injected conjunctivæ.

b. Nervous system. Increased temperature, 100° to 103°, headache, lumbar pains, troubled sleep, special senses often hyperæsthetic.

c. Glandular system and alimentary tract. Secretions diminished and scanty; dark colored irritating urine of high specific gravity. Anorexia—heavy white or yellowish coating on the tongue. Constipation.

2. *Asthenic inflammatory fever.* The general symptoms are the same as those of the sthenic type, except there is *profound depression* in place of over action, and the patient shortly falls into the *typhoid condition*. *Pulse* feeble, rapid, and compressible. *Temperature* fluctuating from 99° or 100° to 103° or even 105°. *Mental condition* dull and torpid, or delirious and busy. *Tongue* dry, with brown or black coat.

How do you treat inflammation?

Locally and constitutionally.

Give the local treatment of inflammation.

Remove the cause. Rest, either general by putting the patient to bed, or local by the employment of splints and bandages.

Position. Elevation with relaxation of all structures by position.

Cold, may be employed with or without moisture; ice-bag, irrigation, rubber tubes, cold compresses, and evaporating lotions. Use in the *beginning of acute inflammation*.

Heat, may be combined with moisture; hot cans or bottles, poultices, spongio piline, irrigation, baths, douches.

Local depletion. Cups, leeches, and scarification.

Counter-irritation. Tr. iodin., mustard plaster, turpentine, chloroform liniment, actual cautery, seton, issue.

Vesication. Fly blister, cantharidal collodion.

Pressure. Either direct, or on the main bloodvessel of the part.

What are the contraindications to the use of cold in inflammation?

It should not be employed where there is great impairment of vitality, either local or general, where it is disagreeable to the patient, or after inflammation is fully established.

How does heat control inflammation?

It restores tonicity to the bloodvessels, increases the rapidity of the circulation, hastens resolution, and is a powerful vitalizer.

Under what circumstances are heat and moisture indicated?

Where there is great tension; where sloughs or dead parts are to be separated; where suppuration is taking place.

What conditions indicate the employment of local depletion?

A condition of vascular engorgement so great that the vitality of the part is threatened; instance, scarification in prolapsed hemorrhoids or acute conjunctivitis

Describe cupping.

If the blood is to be merely drawn to the surface, dry cupping is employed. This may be accomplished by a regular apparatus, or by lighting a few drops of alcohol poured into a small cup or glass, and suddenly clapping it to the surface to be treated. A powerful vacuum is created, and the skin is drawn far into the hollow of the cup. If blood is actually to be abstracted, *wet cups* are used. Incisions are made through the skin, and free bleeding is encouraged by applying cups over these parts.

Describe leeching.

The Swedish leech is generally used; it draws about f$\bar{3}$ss of blood. Wash the surface of the skin carefully, apply a little milk or blood to it, put the leech in a wide-necked bottle, and press the mouth of the bottle against the surface to be bled. Let the leech *drop* off, and check the bleeding either by a pledget of styptic cotton, by compress and bandage, or by passing a harelip pin through the depth of the leech bite and tying around it.

What parts should be avoided in applying leeches?

Leeches should not be placed over loose cellular tissue. Instance, the eyelids and the scrotum.

When do you use counter-irritation?

As counter-irritation acts by drawing the blood from the inflamed part, it may be used in the very beginning as a means of aborting inflammation. It may be employed for the relief of pain, or, as inflammation is subsiding, its use may materially hasten resolution.

Describe the application of counter-irritants.

A mustard plaster must never be allowed to blister. Mix one part mustard, two parts flour, and cover with a thin film of egg albumen or molasses. The more severe forms of counter-irritation, the actual cautery, the seton, and the issue, are especially applicable to chronic inflammation. In using the actual cautery the part may be previously anæsthetized by freezing. The *seton* is made by passing some strands of silk or other material through a pinched up fold of the skin, and leaving them in place, slightly moving them from day to day to keep up irritation. The *issue* is an ulcer made by cautery or chemicals, and kept open by a foreign body, such as a pea or a pebble.

Describe vesication.

This is really a powerful form of counter-irritation combined with depletion. Cantharides in some of its forms is generally used, either the cerate or cantharidal collodion. After six hours apply a poultice; small blisters frequently repeated are termed *fugitive blisters*.

What dangers attend the use of cantharides?

It may be absorbed and produce strangury, *i. e.*, inflammation of the genito-urinary tract, attended with great pain, and constant straining to pass water, with the evacuation of a few drops at a time. *Treat* by opium and belladonna suppositories, demulcent drinks, warm sitz baths, and leeches. *Avoid* by removing the blister after six hours and applying a poultice, or by incorporating camphor with the cantharidal cerate.

In old and debilitated persons *extensive sloughing* may follow the use of blisters.

When is pressure used?

Either in the very beginning, or after the inflammatory swelling has reached its height. It supports the bloodvessels, prevents exudation, and hastens resolution. The ordinary or the rubber bandage may be employed. Often the sand bag or shot bag is of service.

Give the constitutional treatment of inflammation.

1. *Bleeding or general depletion.* To be employed only in the *strong* and *plethoric* at the *beginning of an attack*, and where *life* or the *vitality* of an *important organ* is threatened by the *violence of the congestive symptoms.* Instance, incipient meningitis or pulmonitis. Place the patient in a *semi-recumbent posture*, tie a cord or bandage about the middle of the arm, making enough tension completely to stop the venous circulation, thoroughly disinfect the skin in the region of incision. Under all antiseptic precautions divide the median cephalic vein, and when sufficient blood has been drawn close the wound with a compress of iodoform gauze; remove the fillet from the arm, apply a small antiseptic dressing, and put on a tight spiral reversed of the upper extremity, carrying the hand in a sling. In

case of brain congestion bleed from the external jugular. *Bleed till the pulse becomes soft and slow.*

2. *Cardiac sedatives.* Used where the pulse is full and bounding in acute inflammation. Tr. aconit. rad. gtt. ij, or tr. verat. vir. gtt. v, hourly. Ex. gelsem. fl. ℥ v every two hours. Carefully watch the effect of cardiac sedatives, especially aconite.

3. *Diaphoretics and diuretics.* Applicable to nearly all forms of inflammatory fever. Liq. ammon. acetat. or mist. pot. cit. f℥ss, spirit. aeth. nit. f℥ss well diluted, or pot. nit. gr. v, every two hours. Citrate of caffein or infusion of digitalis may also be given.

4. *Cathartics.* In the beginning of an acute attack of inflammation the bowels should be thoroughly cleared. This may be effected by blue mass gr. x, followed in six hours by a seidlitz powder, or calomel gr. ⅛. sod. bicarb. gr. iij, repeat every hour till evacuation, or liquorice powder ʒj. Keep the bowels regulated by Janos or Carlsbad water.

5. *Antipyretics.* Quinine gr. xx, antipyrine gr. xv, antifebrine gr. v. Not to be used unless the fever exceeds 103°.

6. *Anodynes.* Morphia for acute pain, gr. ¼ hypodermically. Bromide and caffein for headache. Chloral and bromide for restlessness.

7. *Stimulants.* Always in the asthenic or typhoid form of surgical fever. Where there are symptoms of depression, brandy, whiskey, or wine, given at regular intervals with the food.

8. *Tonics.* After the acute stage has passed, tr. cinch. comp. elix. calisay., or quinine.

9. *Diet.* Water and cracked ice for two or three days if the symptoms are very acute, and the affection not liable to terminate in the typhoid condition. Follow by milk taken in small quantities and at regular intervals. As the fever subsides the diet can be rapidly increased. For adynamic fever fullest diet the patient can digest from the first. Milk three pints daily with malt, oyster juice, raw oysters, peptonized raw-meat juice, liquid peptonoids, beef tea, etc.

What symptoms call for the use of stimulants?

A weak pulse, particularly if associated with *delirium*. The guide as to the quantity to be employed is the *pulse*; if it becomes *slower and fuller* the stimulants are doing good.

What are the symptoms of chronic inflammation?
The same as in acute but less marked; any or all of the cardinal symptoms may be so slight as to escape notice.

What are the causes of chronic inflammation?
Preceding acute inflammation. Long continued local irritation or functional activity. Constitutional weakness or diathesis.

What is the pathology of chronic inflammation?
A large amount of plastic lymph is effused and undergoes partial organization, causing considerable induration. This induration greatly slows the circulation by compressing the blood-vessels. The infiltrated tissues undergo fatty degeneration and may break down forming *cold abscesses*.

How do you treat chronic inflammation?
1. *Local. Remove cause.*—May be sequestrum or foreign body. *Rest*, general, in bed; local, by splints and bandages. *Local depletion.*—By leeches and scarification. *Vesication.*—Small and frequently repeated blisters. *Counter-irritation.*—Actual cautery, setons, issues. *Alteratives.*—Tr. iodin., unguent. iodin. comp., unguent. hydrarg. cum belladon. *Irrigation and pressure.*—Apply a tight roller bandage and keep wet by cold or hot irrigation. This is the most efficient local treatment of chronic inflammation. *Massage—Electricity.*

2. *Constitutional.*—Fresh air, generous diet, stimulants, and tonics. Mercury, iodine and iodides. Cod-liver oil.

When must mercury be avoided?
In strumous, tubercular, and broken down constitutions.

How is mercury given?
Hydrarg. chlor. mit. gr. ⅛, Dover's powder gr. ij., give every two hours. Mainly used in head injuries or inflammations, also advised in inflammation of all serous membranes.

What is meant by salivation?
The constitutional effect of an overdose of mercury. Early

symptoms, a *fœtid breath* followed by *tenderness of the gums*, noticed on chewing. Metallic taste in the mouth. Copious flow of saliva.

How do you treat salivation?
Stop the mercury, open the bowels, use a mouth wash containing tr. myrrh. and pot. chlor. Administer belladonna or atropia in fairly full doses.

Abscess.

What is an abscess?
A collection of pus surrounded by a wall of lymph. An abscess is a hollow ulcer.

Describe the formation of abscess.
From excessive or continued irritation there is an exudation so copious that not only are the lymph channels blocked but there is absolute blood stasis and coagulation; the central portion of the exudation and the involved tissues perish forming pus; the peripheral portions, however, are not absolutely cut off from nutrient blood, they undergo organization and form around the central part a bank of organized lymph or granulation tissue; this serves a double purpose; to prevent the extension of the suppurative process, and to provide for the healing of the abscess when the pus is evacuated. The direct cause of the pus formation is the presence of micro-organisms in the exudate.

What symptoms denote the formation of an abscess?
Throbbing pain. Increase in swelling. Color darker, surface at times glazed. Tendency to point. *Fluctuation* elicited by palpation, percussion, and pressure. *Rigors* and *fever*.

In what direction does an abscess point?
In the direction of least resistance. This is usually, but not always towards the surface.

What local symptoms point to deep suppuration?
Pain and *œdema*.

How do you treat an acute abscess?

Endeavor to abort by the use of heat (110°), cold, local depletion or blisters. If these fail, relieve tension and *hasten suppuration* by poulticing. When fluctuation is detected, *open* under antiseptic precautions, wash out the cavity with bichloride solution 1 : 1000, drain, and apply a Lister dressing.

How do you open an abscess?

If superficial, with one quick cut. If deep, make an incision with the scalpel to the deep fascia, through this a director is passed and forced on till it enters the abscess cavity. A pair of dressing forceps, closed, is carried along the director; by opening these and drawing out forcibly a free opening is made without endangering bloodvessels. In evacuating pus, bear in mind that any violence, which breaks down the organized walls of lymph or granulation, retards healing, hence if pus is squeezed out it must be by means of gentle pressure made with pledgets of cotton.

In what regions must abscesses be opened before fluctuation is detected?

1. Ischiorectal, to prevent pointing into the rectum (path of least resistance). 2. Perineal. 3. Palmar. 4. Tonsillar. 5. Postpharyngeal. 6. Any abscess near important bloodvessels or beneath deep fasciæ.

What circumstances may retard the healing of abscess after incision?

1. *Want of free drainage.* To remedy, enlarge the opening, or make another in a more dependent position, or insert drainage tube.

2. *Imperfect apposition* of the granulation walls, *hemorrhage*, or *break in the limiting walls* allowing an infiltration of pus into the surrounding tissue. Treat by compress and bandage.

3. *Indolent granulations* or *constitutional weakness*. Treat locally by stimulating applications. Cu. sulph. or argent. nit. gr. iv to aq. f ℥j, iodoform; the constitutional condition must be remedied by tonics and stimulants.

How does a chronic abscess differ from an acute one?

The course is slow, the signs and symptoms are slight or wanting. The tendency to point is not marked, pus accumulating at times to an extraordinary extent before the skin shows signs of yielding. The pus corpuscles are broken up and there are few or no micro-organisms to be found on microscopic examination. The granulation wall is very thick, partially organized into connective tissue, and showing little tendency toward the production of healthy granulation. The condition is one of passive congestion rather than active hyperæmia, hence the name *congestive abscess*; called also *cold abscess* from the slight development of inflammatory heat.

What are the constitutional symptoms of chronic abscess?

May be slight or wanting till the abscess bursts or is opened, when *hectic* quickly develops; by this is meant a daily rise in temperature, often preceded by rigors, and followed, after some hours, by profuse sweating with subsidence of fever. Emaciation is continuous and rapid.

How do you treat a chronic abscess?

Generous diet, stimulants, tonics, iodide of iron, and cod-liver oil. Unless the abscess is *stationary*, and giving no trouble either directly or indirectly, open at once under strictest antiseptic precautions. *Aspiration* followed by *pressure* may succeed when there is no bone involvement. Usually *incision* will be necessary; the cut must be as far removed from sources of contagion as possible (hence open psoas abscess *above* Poupart's ligament), and planned to thoroughly drain the cavity. Irrigate daily with 5 per cent. sterilized salt solution, 1 per cent. carbolic, or 1 : 6000 bichloride. Apply each time a complete antiseptic dressing, providing cushions of jute, oakum, sea moss, or cotton to receive and absorb the discharge.

What are the chief characteristics of tubercular abscess?

They are *chronic*, have a tendency to *caseation* and *long-continued discharge*, and affect mainly *bones*, *lymph glands*, and *lungs*.

How do you treat tubercular abscess?

Thoroughly remove the affected area by means of the knife or curette under antiseptic precautions.

What is a residual abscess?

An abscess which appears at or about the seat of a former suppuration; commonly due to caseous masses.

What is a sinus?

A suppurating canal, left by an imperfectly healed wound or abscess.

What is a fistula?

A communication between two mucous cavities, or between a mucous cavity and the external air, by means of a suppurating canal.

How do you treat sinus and fistula?

Remove all irritating causes and bring the walls together by pressure, employing stimulating injections (silver, copper, zinc); or freely lay open, and by gentle packing with iodoform gauze, cause healing from the bottom.

How do you diagnose abscess from aneurism?

Should abscess occur in the immediate neighborhood of a large vessel the *history* will be one of previous inflammation; the pulsation of abscess is a *simple lifting impulse*, not an expansive throb; the abscess may be *absolutely isolated* from the artery by manipulation; pressure on the distal side does *not increase* the *tension* of abscess, nor does pressure on the proximal side diminish it. Abscess gives *no thrill, no bruit;* finally, if there be the chance of a doubt, the *exploring needle* gives pus from the abscess.

How do you distinguish encephaloid disease from abscess?

In soft cancer the course is chronic, and at first painless; it presents multiple eminences, has large purple veins coursing over it, and is elastic rather than fluctuating.

Ulceration.

What is ulceration?
The molecular death of tissues, leaving a solution of continuity, and accompanied by a discharge.

What are the causes of ulceration?
1. Predisposing, *quantity* and *quality* of the blood, together with the *freedom* and *rapidity* of the circulation.
2. Exciting, *irritation*, physical or chemical.

What is the pathology of ulceration?
As for abscess; from over-crowding, the tissues and effused matter about the focus of inflammation perish, the peripheral areas become vascularized, and are converted to granulations.

What is a granulation?
A capillary loop about which are clustered leucocytes, held together by a slight amount of intercellular material.

Describe healthy granulations.
Cherry-red, non-sensitive, elastic, and discharging laudable pus.

By what processes is ulceration healed?
By *granulation* and *cicatrization*. While the dead central parts of the ulcer come away as a thin discharge called *ichor*, the exudation beneath and around is becoming vascularized, capillary loops shoot out toward the surface (the direction of least resistance); about each loop clings a cluster of living leucocytes, and a surface of healthy *granulation* is established, discharging laudable pus. Cicatrization now begins, the surrounding skin sinks to the level of the granulations, and its epithelial cells undergo segmentation and grow as a ring about the periphery toward the centre of the ulcer; this skinning over is denoted by a blue film, and while it is extending the ulcer is contracting, from conversion of leucocyte to fibrous tissue; this contraction goes on long after the ulcer is entirely healed, and may cause great deformity. The process of skinning and contraction is called *cicatrization*, the result is a *cicatrix* or scar.

Describe a cicatrix.

At first *blue*, it finally becomes *white*, the progressive contraction of the connective tissue squeezing all the blood from the part. A cicatrix has neither nerves, glands, lymphatics, nor hair; it readily ulcerates, and is slow in healing.

What is an ulcer?

A surface of granulations.

Name the varieties of ulcers.

1. Local.
    a. *Simple* healthy or healing.
    b. *Complicated* or spreading.
2. Constitutional, *strumous, syphilitic*.

Of the *complicated or spreading* we have the *fungous*, the *œdematous*, the *inflamed*, the *sloughing*, the *phagedenic*, the *indolent ulcers*.

Describe a simple or healthy ulcer.

Granulations, healthy, cherry-red, small, uniform, not painful. *Discharge*, laudable pus in small quantity; if the ulcer has been treated antiseptically the discharge is serum. *Shape*, oval, regular. *Edges*, gently sloping, moderately indurated, showing the blue line of beginning skinning. *Surrounding skin* soft and flexible.

Give the treatment of simple ulcer.

In the forming stage, abort or limit by rest, elevation, local depletion, and cold; at the same time treating the rather high constitutional symptoms by withholding food, giving abundance of water, iced drinks, or cracked ice, opening the bowels, and, if necessary, administering morphia hypodermically to control the pain.

When disintegration is evident hasten the separation of the dead from the living tissues by warm antiseptic poultices (sponges, lint, or gauze soaked in weak bichloride solution 1:6000, and covered in by waxed paper and a bandage). Milk diet. *When the dead part is separated* leaving a surface of healthy granulations, cleanse with sterilized salt solution 5 per cent., or very weak antiseptic lotions, bichloride 1 : 10,000. Cover with protective or gutta-percha tissue, and apply a light antiseptic dressing, finishing with moderately firm pressure by a roller bandage. Full diet. A healthy ulcer heals kindly under nearly any dressing.

Describe the inflamed ulcer.

A simple ulcer may become converted to an inflamed ulcer by any of the local or constitutional causes which give rise to inflammation. Instance, debauch, injury, etc. *Granulations*, at first bright red, become dusky, finally break down forming a gray, ragged, sloughing surface. *Discharge*, very profuse, consists of pus and the *débris* of broken down tissue. *Edges*, irregular, deep, sharply cut, indurated. *Surrounding skin*, red and œdematous. *Pain* and *tenderness* acute. Constitutional symptoms well marked.

Give the treatment of inflamed ulcer.

A saline cathartic in the beginning of the attack. Rochelle salts ℥j. Rest in bed with elevation of the part. Local depletion by leeches, or incisions into the edge of the ulcer. Hot antiseptic poultices. Low diet, opium to relieve pain.

Describe the sloughing ulcer.

Very commonly associated with venereal disease. This is but an *aggravated inflamed ulcer*, and is characterized by the *same peculiarities*, with the addition that there is a *rapid spreading* attended by *destruction* of *visible portions* of the tissues which are thrown off as offensive gray sloughs. All symptoms, both local and general, are aggravated.

How do you treat sloughing ulcers?

Treatment on the same lines as for inflamed. Constitutional condition must receive particular attention, as all sloughing processes tend rapidly towards exhaustion. Charcoal or antiseptic poultices till sloughs come away.

Describe the phagedenic ulcer.

This form is an aggravated sloughing ulcer. Found only in venereal disease or in patients with profoundly depressed constitution. The *granulations* are absent, being replaced by gray sloughs; the *discharge* is ichorous, containing shreds of dead tissue; the *edges* are ragged, dusky red, and extensively undermined; the *surrounding skin* œdematous, red. The extension is very rapid, may destroy an entire organ (the penis), and is attended by severe constitutional symptoms of the adynamic type.

Give the treatment of phagedenic ulcer.

Clear the bowels. Rich nourishing diet, stimulants, tonics, opium. Continuous warm baths during the day, with iodoform dressing at night. Or the ulcer may be treated by charcoal poultices and antiseptic washings till sloughs are separated.

Describe the serpiginous ulcer.

This is really a phagedenic ulcer. Its course is *slow* but *persistent*; it may produce most extensive destruction of tissue.

Treatment. Constitutionally, supporting; locally, actual cautery, or as for phagedenic ulcer.

What is an irritable ulcer?

An ulcer which presents the features of an inflamed ulcer, together with great pain, out of all proportion to its apparent cause. This pain is probably due to the stretching of small nerve branches.

Treatment. Subcutaneous section of the nerve branch supplying the ulcerating area, or applications of chloral gr. xx., or argent. nit. gr. xx. to the ounce of water.

What are fungous and œdematous ulcers?

In the *fungous ulcer* the granulations grow above the level of the surrounding skin, and may spread out as a cauliflower or mushroom-like growth; they bleed readily. *Cause*, obstruction to venous return from undue contraction of surrounding tissues.

The *œdematous ulcer* is characterized by large, pale, flabby, watery granulations which have a tendency to become fungous. *Cause*, venous obstruction combined with struma or systemic depression.

How do you treat fungous and œdematous ulcers?

Astringent applications. Powdered alum, glycerole of tannin, followed by *compression* applied by means of imbricated adhesive straps and a tight roller bandage.

Excision. If these means fail, or if the granulations have assumed a mushroom-like growth, shave off level with the surface, dust with iodoform, and apply an antiseptic dressing, with a tight roller bandage over the whole.

Describe the indolent, callous, or chronic ulcer.

Granulations. Never healthy, usually small, scanty, and brickdust-red; frequently fungous or œdematous.

Discharge. Ichorous or sanious pus.

Edges. Everted or inverted, irregular, never gently sloping. Blue line of skinning absent.

Surrounding skin. Discolored, often eczematous and densely indurated.

Occurs. After middle age, and in those whose occupation requires long standing.

Favorite seat. The outer surface of the lower third of the leg, because: 1. It is an exposed portion. 2. There is little cellular tissue separating skin from bone. 3. Its dependent position favors passive congestion and thrombosis.

Course. Exceedingly slow, may last many years.

Constitutional symptoms. None.

The *eczematous* and *varicose* ulcers are simply chronic ulcers with marked development of the affections from which they take their names.

What prevents chronic ulcers from healing?

From long congestion the bank of lymph becomes redundant, and is, in part, converted to imperfect fibrous tissue, which, by pressure upon the vessels, blocks the circulation.

How do you treat chronic ulcers?

Cause the absorption of the obstructing bank of lymph. Healing granulations will then appear. This is accomplished by *heat, moisture,* and *pressure.*

Treatment. Soak the ulcer for two hours at night in warm 4 per cent. boracic acid solution, followed by a thick poultice (boracic acid solution and ground flaxseed, the surface being coated with boracic ointment), well protected by oiled silk, or waxed paper, so that it may not cake before being removed. In the morning, substitute for the poultice strips of lint wet in boracic lotion, and imbricated over the affected region; cover these strips with waxed paper, and apply very carefully over the whole a roller bandage, taking in the foot and going as high as the knee: at night remove the dressing and soak again. Continue this treatment for three or four days, or until the bank of induration is softened, then *strap*. Use adhesive plasters cut in strips one inch wide, and long enough to extend nearly around the limb. After elevating the leg and allowing the blood to drain out, begin the dressing by applying the first strap two inches below the lower border of the ulcer, making firm pressure as it is carried around the leg or foot; the next strap is applied nearer the ulcer, overlapping the first for two-thirds of its width; so continue till the ulcer is reached, when the straps must overlap as before, but in applying them, first fasten one end, then press the edges of the ulcer together, diminishing its size as much as possible, and secure it in this position by continuing the strap firmly across it and around the limb. The straps must entirely cover in the ulcer and an area two inches above and below. Over the straps apply a layer of lint, and cover in the whole by a closely fitting roller bandage. The dressing is removed and reapplied as required by the amount of discharge. If this method cannot be carried out, apply a *Martin's rubber bandage* directly to the skin; removing at night; wash night and morning in boracic lotion.

A *blister* applied to the entire ulcer and surrounding skin may cause the induration to disappear. *Incisions*, or *shaving off* of the induration may be required.

What are the characteristics of strumous ulcers?

Favorite seats *neck* and *groin*. Chronic, painless, discharge a thick oily pus, granulations œdematous, skin extensively undermined, and overhanging the ulcer in the form of loose blue flaps.

What ulcers are mostly found on the leg?

Varicose, traumatic, and syphilitic. A non-traumatic ulcer of the *upper third* of the leg is mostly *syphilitic*.

What ulcers chiefly affect the face?

Rodent ulcers, and those due to lupus, syphilis, or epithelioma. The *rodent ulcer* is distinguished from the *epitheliomatous* from the fact that it does not involve lymphatic glands, nor induce secondary deposits; its course is very slow; its base is smooth and glossy, with little or no discharge; its edges moderately indurated, smooth, round, and rolled over.

Describe skin grafting.

By skin grafting is meant the placing on granulating surfaces of healthy epidermis for the purpose of hastening cicatrization and preventing subsequent contractions. It is chiefly applicable where the granulating surface is large, or conspicuously placed, or slow in healing. The granulations must be healthy, discharging very slightly, and preferably aseptic. This may be accomplished by washing with weak bichloride solutions and dressing antiseptically for several days before the operation. The area from which the grafts are taken should be thoroughly washed with soap, water, and bichloride, 1 : 1000, followed by 5 per cent. sterilized salt solution (sodium chloride 5 parts, water 95 parts, boil for one hour). By means of a scalpel, scissors, or a razor, small or large pieces of cuticle, including the rete mucosum, but not the corium, are removed, and placed, fresh surface down, on the granulations, from which all antiseptics have previously been washed by liberal salt solution irrigations.

Apply protective wet in salt solution, and either a sterile, or an antiseptic dressing, covering in the whole with a tight roller bandage. By this method strips of skin, $\frac{1}{4}$ in. by 2 in., may be transplanted and retain their vitality. The grafts should be taken from young healthy persons.

Mortification.

What is mortification or gangrene?

Death in mass.

What is a slough or sphacelus?

That portion of tissue affected by mortification.

What are the causes of gangrene?

1. *Direct violence* from physical or chemical agencies.
2. *Deficient blood supply* from inflammatory engorgement, weak circulation, diseased vessels, embolus, or thrombus.

Name the two commonest forms of gangrene.

1. Acute or moist. 2. Chronic or dry.

What structures resist gangrene?

Arteries (hence thrombi form before their walls are disintegrated, and bleeding is prevented), nerves, tendons, and bones.

How is gangrene limited?

By a reactive inflammation. A wall of granulation is thrown out, at the expense of the healthy tissues, by which the slough is separated from the living parts.

What first indicates the limit of gangrenous processes?

The line of demarcation. A red line due to capillary congestion, indicating the beginning of inflammatory reaction.

What follows the line of demarcation?

The line of separation. A line of ulceration or granulation.

What are the general indications in the treatment of all gangrenous processes?

Keep the dead or dying part *thoroughly aseptic.* Cleanse, disinfect, and wrap in thick layers of antiseptic wool, cotton, or gauze. Carefully guard against the *invariable tendency* to *adynamia.*

What are the symptoms of acute mortification?

Synonym: *Local traumatic gangrene.*

Usually acute inflammatory symptoms with evidence of great local congestion, and intense burning pain. The pain ceases, there is loss of sensation, of power to move the part. The temperature falls, and pulsation of the arteries cannot be detected. The color, at first dusky-red, turns to blue, to purple, to dirty brown, or black. Blebs form, the course of the superficial vessel is marked by lines of dark discoloration. Even yet vitality may be restored. If, however, the cuticle separates from the derm and can be rubbed off by light pressure, if there is crackling, emphysema, and foul odor, death is absolute.

The constitutional symptoms are those of inflammatory fever, but of an *adynamic* or *typhoid* type. Rapid, feeble pulse, low delirium, etc.

How do you treat acute mortification?

Preventive. *Relieve tension.* Remove tight bandages. Evacuate retained discharges. *Freely incise* inflammatory congestions. Massage. Render the part aseptic; wrap in antiseptic wool.

If the slough is thoroughly established, and is putrid, charcoal poultices or wet bichloride dressings may be used; otherwise, dry antiseptic dressings are indicated.

Amputate when the *line of demarcation is formed.* (In the hand and foot spontaneous amputation generally gives a better stump than the surgeon's knife.)

Constitutional treatment: Very free stimulation, full nourishing diet, quinine, and opium.

What is spreading traumatic gangrene?

An acute, rapidly spreading, moist gangrene, dependent on a specific micro-organism. It appears shortly after severe traumatism, and before the line of separation can form, extensively invades the tissues, and causes death from exhaustion or septic poisoning. All local inflammatory symptoms may be absent: swelling, discoloration, and loss of temperature circulation and sensation, denoting the extension of the process. In other cases, an inflammatory redness and induration precede the gangrene. The constitutional symptoms are *profoundly adynamic.*

How do you treat spreading traumatic gangrene?

Immediate amputation through *healthy tissue*.

What is hospital gangrene?

An epidemic, contagious, gangrenous process, dependent upon the presence of micro-organisms, which destroys granulations, attacks the tissues lying about and beneath them, and rapidly produces extensive sloughs.

Give the symptoms of hospital gangrene.

As for acute mortification. The surface of a wound, or its margins, are rapidly converted into an extensive slough, there is surrounding œdema and congestion, the discharge is foul, the process rapidly extends.

The constitutional symptoms are adynamic; high temperature at first, with weak, quick irregular pulse, wet surface, and, frequently, muttering delirium.

What circumstances predispose to attacks of hospital gangrene?

Over-crowding, deficient ventilation, want of proper nourishment, or any depressing cause.

How do you treat hospital gangrene?

Isolate the patient. Break up the sloughs by thrusting closed dressing forceps through them, and withdrawing the forceps opened. In these openings make a thorough application of pure bromine, nitric acid, or other escharotic. Dress with antiseptic charcoal poultice, and subsequently observe the most rigid asepsis in regard to wound treatment.

Constitutionally give *stimulants, free diet*, quinine, iron, and opium.

What is cancrum oris?

Synonym. Gangrenous stomatitis.

It is a gangrenous ulcer of the cheek or gums, occurring in poorly nourished children. It is frequently developed after an attack of measles, scarlet fever, or typhoid fever. It usually appears opposite a rough or decayed tooth, which has caused an abrasion. It is seen in the mouth as an offensive, sloughing, punched out ulcer; on the external surface of the cheek as a glazed, dusky red, indurated spot, which is shortly converted into a black slough, causing perforation, and extensive destruction of tissue. The constitutional symptoms are those characteristic of all gangrenous processes.

How do you treat cancrum oris?

Thoroughly cauterize with nitric acid. Wash at intervals with boracic acid lotion, or tr. myrrh. Give internally stimulants, rich milk in abundance, malt, iron, and quinine.

What is noma pudendi?

A gangrenous process similar to cancrum oris, attacking the genitals of female children. *Treatment.* As for cancrum oris.

What is a bed sore?

A sloughing ulcer, due to pressure, appearing on the bony prominences of the weak and badly nourished.

How do you treat bed sores?

Clear away the slough by charcoal poultices, wash and dress antiseptically, relieve the part from pressure by pads, pillows, or air cushions.

Describe a furuncle.

Synonym. *Boil.*

Definition. A circumscribed inflammation of the skin and subcutaneous tissue, terminating in suppuration, and the formation of a central slough or core.

Occurs. In crops, on the neck, nates, and back of the young.

Causes. Systemic depression, and the rubbing into the ducts or hair follicles of the skin of a micro-organism.

Begins as a red pimple, usually with a hair in the centre, increases rapidly in size, causing a purple-red, very painful swelling which may undergo resolution (blind boil), or open, discharging the central core.

Treatment. 1. Pull out the central hair, wash thoroughly with bichloride, apply 50 per cent. ichthyol ointment. 2. Inject with ℥v. of a 10 per cent. solution of carbolic acid. 3. If inflammatory symptoms increase in severity, apply spongio piline dipped in hot boracic or carbolic acid lotion. 4. When fluctuation is evident, incise, syringe the cavity with antiseptic solution, and apply an antiseptic dressing, making firm pressure.

What is a carbuncle?

An inflammation of the skin and subcutaneous tissues, involving a much larger surface than furuncle, and attended by the formation of sloughs of considerable size.

It differs from boil in being *much larger, flattened* instead of conical, and accompanied by great surrounding *œdema*. The skin gives way in *several places*, sloughs of some size are discharged. Constitutional symptoms are *severe*.

Occurs in the aged and debilitated.

Cause. The rubbing in, by friction, of a micro-organism.

Seats. Neck, back, nates. When occurring on the face or head it is exceedingly fatal.

Give the symptoms of carbuncle.

A hard, brawny, flattened, dusky-red area of induration, circular in shape, and riddled with apertures, through which a gray slough can be seen. The constitutional symptoms are severe and of an *adynamic type*.

Give the treatment of carbuncle.

The constitutional treatment should be conducted on the plan indicated for all gangrenous processes. Stimulants, full diet, iron, quinine, and opium may be given. Locally, the affection may be treated by—

1. Tight concentric strapping, leaving a central aperture for the escape of sloughs.

2. Hot fomentations or poultices, the moisture being supplied by boracic or carbolic acid solution. Heat and moisture may be combined with strapping.

3. Injections through the inflamed area, and about its periphery, of carbolic acid (5 or 10 per cent. in glycerine); as much as a half drachm may be used.

4. Crucial incision, and removal by curetting of all the involved cellular tissue. The operation must be done antiseptically. Pack the wound with iodoform gauze, and apply a thick antiseptic dressing.

What is the usual cause of dry gangrene?

Synonym: *Senile* or *chronic gangrene*.

Cause. Arterial obstruction from atheroma and thrombosis.

What are the premonitory symptoms of senile gangrene?

The limb feels *cold* and *numb; tingles* and is subject to *shooting* and *violent pains;* steady deterioration in health.

What symptoms denote the onset of the disease?

The appearance of a black spot, usually to the inner side of the great toe, surrounded by a dusky-red areola, and causing an intense burning pain. There is a slow extension till the entire foot becomes hard, dry, black, and mummified.

How do you treat dry gangrene?

Disinfect the part and wrap in antiseptic wool or cotton. Allow a generous diet. Give tonics and stimulants; opium two or three grains daily.

Under what circumstances is amputation required in gangrene?

When the line of separation is formed.

Immediately, in spreading or traumatic gangrene.

In gangrene from arterial occlusion, when the seat of the occlusion can be certainly determined. Instance, wound or ligation of an artery.

In senile gangrene, only when the *line of separation has formed*, and exploratory incision shows that the *arteries above are healthy*.

WOUNDS.

The Germ Theory.

Outline the germ theory of putrefaction.

Putrefaction is the result of the growth of micro-organisms in the substance which putrefies. These micro-organisms are divided into—

1. Non-pathogenic, or those which do not *directly* create disease.

2. Pathogenic, or disease creating.

Among the non-pathogenic, are included those which can live or grow only in dead or dying matter, termed *saphrophytic*. These saphrophytic micro-organisms, entering a wound in which there is much pent-up discharge and dying tissue, rapidly increase, and produce certain irritating substances, called *ptomaines*. The absorption of ptomaines into the system gives rise to the symptoms which are characterized as *septic intoxication, ptomaine fever, sapræmia,* or *septicæmia.*

Pathogenic micro-organisms thrive not only on dead matter, but invade and destroy the living tissues. They may be carried through the circulation to all parts of the body, increasing with incredible rapidity wherever deposited, destroying tissue, and forming fresh centres for the production of poisonous products. They enter the system, by a process of direct inoculation, through wounds. Nearly all pathogenic microbes are either micrococci (spherical) or bacilli (rod-shaped).

What are the general principles of antiseptic treatment?

1. Prevent putrefaction. 2. If it has already occurred, stop its further progress.

Since putrefaction depends upon the presence of an organism, and a soil in which it can grow, the indications for the prevention of this process are—

1. *Exclude all organisms* from the wound. This may be accomplished by most minute attention to the details of surgical cleanliness.

2. *Remove organisms* from the wound, before they can work harm, by irrigation.

3. *Destroy organisms*, by bichloride or other germicides.

4. *Remove the soil* in which organisms can flourish, by free drainage.

5. *Prevent the formation of favorable soil*, by avoiding tension or unnecessary manipulation, and by careful dry dressing.

What is the distinction between antiseptic and aseptic?

Aseptic means *germ free; antiseptic* means *germ destroying*. The surgeon who does not practise antisepsis cannot procure asepsis. An aseptic wound is the result of antiseptic treatment. Dressings sterilized by heat have undergone as thorough antiseptic treatment as those saturated with bichloride. By an aseptic dressing is meant the application of substances *previously sterilized*, but containing, at the time of application, no germ-destroying agents. Antiseptic dressings contain germ destroying agents.

Shock.

What is shock?

A lowering of the vital powers consequent on profound mental or physical impression.

Shock is a vaso-motor paralysis, affecting also the heart, and chiefly the abdominal vessels.

What are the causes of shock?

1. *Powerful mental impressions*, joy, grief, and fear.

2. *Mechanical injury;* traumatism, especially of the abdomen; burns, scalds, cold; gunshot, lacerated, and contused wounds. As predisposing causes can be classed all conditions which cause enfeeblement of the resisting powers. Instance, Bright's disease, sedentary occupation, and hemorrhage.

What are the symptoms of shock?

Pulse first slow, then rapid, feeble, compressible, and scarcely perceptible.

Temperature sub-normal.

Surface cold, pale, and wet.

Muscular system relaxed, contractility of sphincters lost. Patient lies in any position in which he may be placed. Decubitus usually dorsal.

Nausea and vomiting frequently present.

Consciousness and special senses blunted.

What is your prognosis in shock?

Bad if the temperature falls below 96°, or if reaction is delayed twenty-four hours.

What becomes of a patient suffering from shock?

He either *collapses* and dies from syncope or asthenia, or *reacts*.

Describe reaction.

Healthy reaction is characterized by an increase in the force, and a diminution in the rapidity of the heart's beat, a *rise of temperature*, a restoration of color to the blanched surface, and disappearance of all the characteristics of shock. In other cases reaction may take the form of an *acute fever*, with flushed face, injected conjunctivæ, high temperature, restlessness, jactitation, *active or muttering delirium*, and a full, throbbing pulse. The pulse, however, is *soft and compressible;* the tongue is *dry and tremulous; the symptoms are asthenic*, and are liable to lapse again into profound and fatal shock. This condition is termed *traumatic delirium*, and is a condition of under reaction from shock.

How do you treat shock?

External warmth most important of all treatment. Hot bath, hot bricks or bottles applied along the spine, to the epigastrium, and about the patient's body and limbs.

Position. Dorsal decubitus with head low.

Medication. Atropia gr. $\frac{1}{100}$ and brandy ʒjs, every thirty minutes hypodermically; morphia gr. $\frac{1}{4}$ if there is great pain. Avoid medication by the stomach till reaction begins, as there is no absorption. Hot coffee, or hot, highly seasoned beef tea may be given in small doses by the mouth. When reaction has fairly set in, *stop stimulating*.

Describe the forms of traumatic delirium.

In addition to the form described as an imperfect reaction

from shock, there is an *inflammatory*, a *nervous*, and an *alcoholic* traumatic delirium.

The inflammatory form is characterized by fever and *sthenic symptoms* with either *sthenic* or *asthenic condition*. It develops in from three to five days after the injury, and is really a symptom of septic inflammatory fever. *Treat* as for the fever, applying an ice cap to the head.

The *nervous* and *alcoholic forms of traumatic delirium* have the same busy asthenic delirium, the soft, full, quick pulse, the tremulousness, and absence of fever, the difference being that the *nervous* is not caused by alcohol.

Treatment. Stimulants, bromide, chloral, morphia. Clear the bowels, give plenty of nourishing liquid food highly seasoned.

What is secondary shock?

Symptoms coming on at varying times from the primary shock, and causing death from *heart clot*, are characterized as secondary shock.

Should you operate during shock?

Not unless it is for the relief of a condition causing, or keeping up the shock. Instance, a strangulated hernia, a bleeding artery, a depressed fracture of the skull. The rule is to *wait for reaction*.

Wound Fever.

What is traumatic fever?

Fever following traumatism.

Several forms may develop, the first of which is the *reactive fever*; this *follows the shock* of traumatism. It develops a few hours after an injury or operation, and subsides in one or two days at most. This is the only form of wound fever which should develop in antiseptic surgery.

How do other forms of traumatic fever develop?

Given a wound in which there is tension, or irritation from other causes, or in which a few non-pathogenic microbes are found, there will be a slight amount of inflammation and absorption, and the patient will probably develop *inflammatory fever*, appearing on the second or third day, and lasting from two to six days. Should the wound contain a large quantity of discharge to which micro-organisms have had free access, *septicæmia* or septic intoxication, from the absorption of ptomaines. is developed. If the micro-organisms are allowed to multiply till they overwhelm the tissues and enter the blood current, *pyæmia*, or septic fever attended with the formation of metastatic abscesses, is developed.

What are the symptoms of traumatic inflammatory fever?

Full, strong, rapid pulse, increased temperature ($100°$-$103°$), restlessness, headache, and at times delirium, diminished secretions, coated tongue, anorexia, and constipation.

How do you treat inflammatory fever?

Free the wound from all tension. Provide against the *possibility* of discharge being retained, irrigate thoroughly with $1:1000$ bichloride solution, dust liberally with iodoform, and apply a thorough antiseptic dressing. renewing the dressing daily till the fever subsides. Clear the bowels; give aconite, bromide, or morphia, as required by the symptoms.

What is septicæmia?

A septic intoxication, caused by the absorption of the products of putrefaction. Hence, it is most liable to occur in wounds not treated antiseptically, or in those which, from their depth, extent, or location, cannot be thoroughly disinfected and protected. Instance, compound fractures, wounds involving the peritoneum.

Give the symptoms of septicæmia.

Inflammatory fever may run into septicæmia, or this affection may develop very shortly after the infliction of a wound.

Temperature. Rises suddenly, and is at first very high ($104°$-$106°$), may shortly sink to normal or below.

Pulse. Soft, rapid, and compressible, becoming weak and thready.

Respirations, rapid and shallow.

Nervous condition, heavy, apathetic, somnolent. Rarely, active delirium.

Tongue, dry, hard, and discolored. Teeth covered with sordes. At times *profuse diarrhœa*. Urine and fæces passed involuntarily. Death in collapse.

The wound is always unhealthy, frequently sloughing.

The septic poisoning may be so slight in amount as to cause scarcely recognizable symptoms, or may, within twenty-four hours of the infliction of an injury, overwhelm the system.

How do you treat septicæmia?

Remove the septic matter, and make the wound sterile by irrigation, or continuous baths with bichloride solution. Eliminate the ptomaines by a saline purge. *Support the strength* by stimulants, quinine in tonic doses, nutritious food given frequently in small quantities; milk and malt, peptonoids, raw beef juice. Reduce high temperature by antipyrine, gr. x.–xv., or quinine, gr. xx. Secure plenty of fresh air and sunlight.

What is pyæmia?

A septic fever, characterized by the formation of metastatic abscesses. Pathogenic organisms (staphylococci and streptococci) invade the blood, and are carried from the infected area to all parts of the body, where they are lodged as emboli, and form new foci of suppuration and infection.

What is the difference between traumatic inflammatory fever, septicæmia, and pyæmia?

Simply a difference of degree. They all depend upon the *same cause*, and are of the *same nature*. They occur only in infected wounds, and are due to the septic action of micro-organisms and their products.

What are the symptoms of pyæmia?

Irregularly recurring attacks, characterized by a *marked* and *prolonged chill*, associated with high temperature (104°–106°); followed by a *brief hot stage*, the patient manifesting the symptoms and signs of fever; terminating in a *drenching sweat*, the temperature quickly falling to normal or below. Several such attacks may occur in a day. The strength rapidly fails; the pulse becomes weak and rapid; the tongue dry and brown coated; breath mawkish; *metastatic abscesses are detected in the lungs;* the wound is unhealthy, the discharges ichorous.

How do you treat pyæmia?

Thoroughly cleanse the original source of infection by irrigation, curetting, and antiseptic dressing; if this be impracticable, as in osteomyelitis, amputate. Open and drain all accessible abscesses. Push stimulants to their fullest extent, give quinine in heroic doses (gr. lx. daily), milk and pressed beef-juice in small quantities frequently repeated. Provide for sun-light, and open air.

What is hectic fever?

A continued remittent fever, due to septic absorption; characterized by rigors and *fever* during the afternoon and evening, followed by profuse *sweats* and defervescence during the night. The *pulse is constantly rapid*, the eye bright, the cheek flushed, the tongue red and dry at the edges, the *emaciation progressive*. Instance, the fever of consumption.

How do you treat hectic?

Remove the source of septic absorption, by resection, if it is an infected bone area; by incision and curetting, if it is an abscess. Give tonics, stimulants, and a full nourishing diet. Change of air is beneficial.

Erysipelas.

What is erysipelas?

An infective spreading inflammation, attacking either the skin, cellular tissue, mucous, or serous membranes.

What are the causes of erysipelas?

Predisposing. *Wounds,* particularly those which are septic, together with any local or systemic condition depressing to the vital resistance. Instance, *kidney disease,* intemperance, overcrowding, starvation. Exciting. *Micro-organisms* and their products.

Name the varieties of erysipelas.

1. Cutaneous or *simple*. 2. Cellulo-cutaneous or *phlegmonous*. 3. Cellular or *diffuse cellulitis*.

Describe simple erysipelas.

Constitutional symptoms. Rigors, headache, and fever, the temperature suddenly rising to 103° or 104°; with nausea and vomiting. The fever shortly assumes a typhoid type.

Local symptoms. A rash, rapidly spreading from a scratch, abrasion, or wound, and characterized by *well defined margins*, *rosy-red hue*, *smooth, glazed, œdematous, slightly raised surface*, *stiffness* and *burning pain*, frequently *blebs* or *vesicles*, *involvement of nearest lymphatic glands*. The eruption may suddenly disappear from one part to reappear in another, *erysipelas ambulans*.

The pathogenic organism of simple erysipelas has been isolated. It is found blocking the lymph vessels and spaces *in the spreading borders* of the inflammation, shows up well in dry cover glass preparations, appearing as micrococci grouped in chains, and is diagnostic of erysipelas. The eruption lasts about four days in one part, and as it subsides is followed by desquamation.

Give the treatment of simple erysipelas.

If there is a distinct wound, thoroughly cleanse and drain it. Freely open the bowels by a saline cathartic. Milk diet for the first few days. Tr. fer. chlor. ℞ xx. every two hours from the first: shortly begin quinine, in tonic doses (gr. v. to x. daily), stimulants, and as free a diet as the stomach will bear.

To the eruption apply starch and zinc oxide, equal parts of each, and cover in with cotton-wool; or apply a 50 per cent. ichthyol ointment, over which is placed salicylated cotton.

Describe phlegmonous erysipelas.

The skin and subcutaneous tissues are both affected; the symptoms are, in general, the same as for simple erysipelas, but more marked. The swelling is greater, the edges not so sharply circumscribed, the color darker, blebs and vesicles are more common.

The surface, at first densely indurated, becomes boggy in spots and may break down, exposing extensive sloughs. The constitutional symptoms are well marked, running shortly into the typhoid type. The patient may perish from pneumonia, blood poisoning, or exhaustion.

How do you treat phlegmonous erysipelas?

Constitutionally, as in the case of simple erysipelas. A purge, light milk diet; followed in a day or two by full nourishment, tonics, and stimulants. Iron as before.

Locally. Applications of heat and moisture (hot antiseptic fomentations). *Multiple incisions* as soon as the part becomes brawny, going down to, but *not through* the deep fascia. Check hemorrhage by packing with iodoform gauze. Strict antiseptic dressing.

Describe cellular erysipelas, or diffuse cellulitis.

This is a spreading infective inflammation, which may involve the cellular tissues of any part of the body. Instance, the intermuscular planes, the pelvic cellular tissues.

The constitutional symptoms are the same as those characterizing phlegmonous erysipelas; the typhoid condition appears more quickly, and septic poisoning is more commonly developed.

The local symptoms are at first less marked than in any of the varieties of erysipelas. There is *dense induration* succeeded by *bogginess* and ending in *extensive sloughing*.

Treatment as for cellulo-cutaneous. *Incisions early*. *Stimulating* and *supporting treatment* from the first.

Tetanus.

What is tetanus?

A tonic spasm of the voluntary muscles with clonic exacerbations, due to the introduction into the system of an infective poison.

What are the causes of tetanus?

1. *Predisposing*. Hot climate, exposure to cold and damp, or sudden change of temperature, negro race, lacerated and punctured wounds, burns, frost-bites, all septic wounds.

2. *Exciting*. A micro-organism.

What are the symptoms of tetanus?

A slight stiffness of the muscles of the neck and jaws, with increase of pain, and the appearance of a sanious or ichorous discharge in the wounded part, denote the onset of the disease. All the voluntary muscles, including those of respiration, may become involved. There is intense præcordial pain from tonic spasm of the diaphragm, the countenance exhibits a peculiar grinning expression (risus sardonicus), and at the slightest irritation, such as a breath of air, a loud noise, or an attempt to swallow, violent spasms occur which may variously contort the body. If the spinal muscles are chiefly affected, we have *opisthotonos*, or arching backward, the body being supported on the head and heels. *Emprosthotonos* may be developed, the body being bent forward and rolled up like a ball. More rarely *pleurosthotonos*, or drawing of the body to one side, is seen. The skin is wet, the bowels confined, the temperature about normal; it may rise to 108° or 110° shortly before death. Intellect clear.

What is the prognosis of tetanus?

Bad in acute cases; becomes more favorable if life be prolonged till the twelfth day. Death occurs from spasm of the glottis or respiratory muscles, from syncope, from exhaustion.

What are the diagnostic points of tetanus?

The *absence of fever* from the first, the *tonic character of the spasm*, the *early involvement* of the neck and jaw, the *marked convulsive attacks*, and the *clear mind*.

Give the treatment of tetanus.

Local. Make the wound aseptic. Amputation, nerve cutting, or nerve stretching, have also been advised.

Constitutional. Bromide of potassium up to its constitutional effect (40 to 80 grains every two hours), chloral at night to produce sleep. Morphia may be given; it must be pushed to the extreme limit of safety. To prevent death from asphyxia give chloroform during the spasm. Stimulants, and nourishing diet are indicated from the first.

Hydrophobia.

What is hydrophobia?

A disease due to a specific poison introduced into the system by the bite of a rabid animal.

What bites are especially liable to be followed by hydrophobia?

Those on the face, or involving parts of the body unprotected by clothing.

What is the period of incubation?

It varies from six weeks to three months; it may be a very few days, or many years.

What are the symptoms of hydrophobia?

First stage, or stage of melancholia, itching, burning, or inflammation of the cicatrized wound; anxiety, melancholia, or change of disposition; slight difficulty in swallowing, or a catch in the respiration. After a few days the disease is fully developed. *The stage of excitement* is characterized by clonic convulsions, involving especially the muscles of respiration and deglutition; by mental disorder similar to that of delirium tremens, with periods of maniacal excitement, and intervals of lucidity. It is followed after some days by *the stage of exhaustion and paralysis.* The muscular system is entirely unresponsive, and the dying patient lies motionless; the mind is often clear at this stage.

How do you treat hydrophobia?

At the time the wound is inflicted, cauterize, at once and thoroughly, by hot iron, nitric acid, or caustic potash. *Suck the wound.* If the wound has cicatrized when seen, excise the cicatrix. Send the patient where he can be inoculated after Pasteur's method with attenuated virus.

When the symptoms are pronounced, morphia, chloral, chloroform to relieve suffering. Pilocarpine gr. $\frac{1}{8}$ hypodermically, repeated frequently. Hot vapor bath.

Glanders.

What is glanders?
An infective disease of horses, dependent on a specific microorganism; communicable to man through wounds, or the mucous membrane. In horses it is called *glanders* when it attacks the nasal mucous membrane, *farcy* when it attacks the lymphatic vessels and glands.

What are the symptoms of glanders?
A *discharge* from the nose, thin, sanious, offensive, purulent, with involvement of the submaxillary glands. A *pustular eruption* resembling smallpox, involving the skin and the mucous membrane of the respiratory and alimentary tracts. *Subcutaneous nodules*, shortly breaking down and forming *foul ulcers*. There is *fever*, which quickly becomes *adynamic*, and death takes place within a week from septicæmia or pyæmia. There is a chronic form of glanders with less marked symptoms, and from which recovery is possible.

How do you treat glanders?
Use antiseptic nose washes (boracic acid or weak bichloride solution). Open abscesses. Pursue from the first a tonic, stimulating, and supporting treatment.

Malignant Pustule.

What is anthrax?
A specific infective disease due to the entrance of a bacillus or its spores into the system. Its starting-point is in a scratch or abrasion. It is found, in this country, mainly among those who handle imported hides or wool.

What are the symptoms of anthrax?
A red, itching pimple, followed shortly by a vesicle attended with well-marked, brawny induration. Sloughing begins at once, and the *anthrax pustule* is formed, characterized by a *dry, central slough, surrounded by a ring of vesicles*, peripheral to which there

is an area of redness, induration, and great œdema. The neighboring lymphatic glands are involved. Fever of an adynamic type develops, and the patient commonly perishes of exhaustion or syncope. *Diagnosis* by examining the contents of the vesicle; bacilli from $\frac{1}{2400}$ to $\frac{1}{1400}$ of an inch in length can be detected by low powers of the microscope.

How do you treat malignant pustule?

Freely excise the pustule, and either cauterize the wound with caustic potash or carbolic acid, or wash thoroughly and repeatedly with 5 per cent. potassium permanganate solution. A stimulant, tonic, and supporting treatment is indicated constitutionally.

The Healing of Wounds.

Describe the process of repair in incised wounds.

Repair takes place in all wounds by the organization of plastic lymph.

If the wound is an incised one, if its surfaces are accurately approximated, if it is not subject to irritation, either mechanical or chemical, the exudation takes place in minimum quantity, the red blood corpuscles of the blood clot are absorbed; in twenty-four hours the surfaces adhere, and in two or three days the thin layer of plastic lymph which binds them together is supplied with vessels; this is called *union by adhesion or by first intention*. Inflammation scarcely passes the first stage; there is simply a little hyperæmia, puffiness, and tenderness about the lips of the wound.

If the wound surfaces are not accurately apposed, if they are subject to irritation, either mechanical, from improper dressing, or chemical, from irritating applications or the products of germ life, the exudation becomes excessive; there is death of tissue, there is suppuration; if tension and other sources of irritation be removed by free discharge, the gap is promptly filled in with organized plastic lymph or *granulations*, and the wound heals by *granulation* or *second intention*.

If healthy granulating surfaces can be brought together and

retained in position, permanent adhesion between them takes place at once. This constitutes union by *secondary adhesion* or *third intention*.

Primary adhesion or first intention. The prompt union of divided surfaces without obvious signs of inflammation.

Adhesion by granulation or second intention. The union of divided surfaces by granulation tissue (organized lymph), attended with evident inflammatory symptoms.

Secondary adhesion or third intention. The union of granulating surfaces. Amputation flaps which fail to unite by primary intention heal in this way.

What circumstances prevent wounds from healing by primary intention?

1. *Want of accurate apposition;* from gaping, from extensive loss of substance, from retained blood or wound secretions, or from foreign body.

2. *Want of proper protection.* There may be undue motion of the part, it may be subject to direct mechanical or chemical violence, it may be exposed to infection from poisonous agents.

3. *Defective nutrition,* either *local,* from bad position or from tension, or *general* from constitutional weakness.

The Treatment of Wounds.

What are the general indications in the treatment of wounds?

1. Arrest hemorrhage.
2. Cleanse, and remove foreign bodies.
3. Provide for drainage.
4. Bring the wounded surfaces in contact, and keep them apposed.
5. Provide for absolute local rest.
6. Prevent putrefaction.

Name the varieties of hemorrhage.

Arterial, Venous, Capillary. Internal or concealed hemorrhage indicates bleeding into one of the cavities of the body. *Extravasation* indicates bleeding into the areolar tissue. Further, hemorrhage may be *primary, intermediate* or *consecutive, secondary.*

What are the characteristics of the different kinds of hemorrhage?

Arterial. Bright red blood jets from the wound. Pressure on the arterial trunk *above* checks the bleeding.

Venous. Dark blood wells from the wound. Pressure on the venous trunk *below* checks the bleeding.

Capillary. The blood oozes from the surface of the wound, and collects as a pool in its deeper parts.

What are the constitutional effects of hemorrhage?

A feeble, fluttering, rapid pulse, finally perceptible in the large arteries only. A cold, blanched, wet surface, with colorless lips, and sighing respiration. Nausea. Frequently, uncontrollable restlessness, a roaring in the ears, darkness before the eyes, and horrible sinking sensations. The patient may suddenly faint. In syncope the heart's action is so feeble that clotting may take place and bleeding be permanently arrested, or, on reaction, the clot may be washed away by the returning blood current and bleeding continue, to end in a return of syncope, in convulsions and death. Or the patient may recover, passing into the condition known as *hemorrhagic fever*, an irritative fever characterized by rise of temperature, extreme restlessness, great thirst, and a quick jerky pulse.

A *sudden violent* hemorrhage is much more liable to produce fatal syncope than a slow continuous one. *Infants* bear the loss of blood very badly.

Describe nature's method of arresting hemorrhage.

1. Contraction and retraction of the vessels.
2. Coagulations of the blood aided, after severe bleeding, by *enfeebled heart action* and *alteration* in the *composition of the blood*.

On cutting an artery the muscular fibres of its midddle coat *contract*, narrowing or closing the lumen and drawing the end of the vessel from its sheath; the cut ends also *retract* from each other, owing to the natural elasticity of the artery. Neither contraction nor retraction can take place *unless the artery is entirely cut across*; hence, *complete* section of a bleeding artery often stops the hemorrhage.

Coagulation is excited by the divided vessel wall, the sheath of the artery, and the air; it presently occludes the opening in the artery, and also fills with clot the space left vacant in the sheath by retraction; this constitutes *external clot*. Coagulation also extends from the mouth of the vessel backward, forming a clot, conical in shape, with its base to the wound, and extending as far as the nearest branch; this constitutes the *internal clot*.

By continued hemorrhage the blood is made *more coagulable*; a clot forms too rapidly to be washed away by the feeble arterial wave. Arrest of hemorrhage from veins is due to *coagulation*. *The permanent arrest of hemorrhage* is effected by the *exudation of plastic lymph*, which takes the place of the clot, the subsequent *organization* of this lymph, and the conversion of the occluded part of the artery into a fibrous cord.

What is the constitutional treatment of hemorrhage?

The patient should be laid flat on his back; if the symptoms are very severe, elevate the foot of his bed and apply an Esmarch's bandage to the legs and arms, thus keeping the blood to the nerve centres. Hot bottles may be applied about the body. In extreme cases resort to transfusion. Ether ℥xxx., morphia gr. ¼, should be given subcutaneously. Place a mustard plaster over the heart. Give injections of hot water and brandy. Hot coffee or beef tea in frequently repeated *small doses* by the mouth, if the stomach is retentive. As the patient recovers stop stimulants. Give milk diet at first, increasing as rapidly as possible. Give iron as soon as the stomach will allow of its use. *In all cases avoid stimulants unless life is directly threatened by cardiac failure.* The use of stimulants is frequently attended by a return of bleeding.

Describe the methods of transfusion.

Blood or saline solutions may be used. It must be introduced warm ($98°-100°$), in sufficient quantity to add strength and volume to the pulse, and must not contain bubbles of air. Transfusion may be *immediate*, the blood being passed directly from the vein of the donor to the patient's circulation; or *mediate*, the blood being first whipped and strained of its fibrin, then injected.

How do you check hemorrhage?

By 1. Position. 2. Cold. 3. Heat. 4. Pressure. 5. Styptics. 6. Cautery. 7. Ligation. 8. Torsion. 9. Acupressure. 10. Forcipressure. 11. Constitutional treatment.

What position favors the checking of hemorrhage?

Elevation of the part and forcible flexion. Flexion bends the artery sharply on itself, and is applicable to wounds of the extremities.

Describe the use of cold as a hæmostatic.

Used only to check bleeding from smaller vessels. It causes *contraction* and *coagulation*. Ice, ice-water as a fine forcible stream directed against the bleeding point.

Describe the use of heat as a hæmostatic.

Used to check general oozing from large surfaces. It causes contraction and coagulation. Apply in the form of large compresses wrung out in *hot* (120°-140°) water.

Describe the use of pressure as a hæmostatic.

A graduated compress and a bandage may be used for the permanent arrest of hemorrhage when other means are not available, or when several vessels are bleeding and there is a firm bone against which to make pressure. Instance, wounds of the palm or of the scalp.

As a temporary means of checking bleeding the finger in the wound is most efficient, the hemorrhage from any accessible artery can be checked in this way. The tourniquet and Esmarch's rubber tube are also of temporary service.

Describe the use of styptics as hæmostatics.

Act by *coagulating the blood*, they also contract the arteries. They must be brought into *immediate contact* with the bleeding vessel. They all interfere with primary union. Use powdered alum, tannin, gallic acid, or persulphate of iron; solutions of the same drugs, especially hot saturated solutions of alum may be employed; alcohol, turpentine, chloroform are also recommended. Chiefly useful in checking bleeding from malignant ulcers, or in inaccessible regions. Styptics should be employed in conjunction with pressure.

Describe the use of the actual cautery as a hæmostatic.

It coagulates the blood, causes contraction of the muscular coat of the artery, and forms an eschar which acts mechanically. If the actual cautery is used, it should not be heated beyond a dull red. Secondary hemorrhage may occur when the eschar separates. Applicable where there is difficulty in placing ligatures. Instance, in operation about the bones of the face. Paquelin's cautery or the galvano cautery should be used.

Describe ligation as a means of arresting hemorrhage.

This is the most important of all hæmostatic agents. By the pressure of the thread, the middle and internal coats are divided, and curl up within the vessel, causing clotting; this clotting extends to the first lateral branch.

If the artery is ligated in its continuity, a conical clot is formed on both the distal and proximal sides of the ligature, with the apex in each case pointing away from the thread.

About the ligature there is deposited a layer of plastic lymph; the internal clot becomes infiltrated with leucocytes and organizes; the ligature, if aseptic, is either absorbed or encysted, and the artery is converted into a fibrous cord. If the ligature is septic, or subject to irritation, it separates by *ulceration;* this separation may be accompanied by secondary hemorrhage.

What precautions are observed in applying a ligature?

It must be aseptic. It should include only the vessel. If applied to an artery in its continuity, a healthy part of the vessel must be selected; a *square knot* should be tied.

Of what should ligatures be made?

Carbolized and chromicized catgut; carbolized silk.

Describe the method of applying torsion as a hæmostatic.

Torsion consists in seizing the artery in torsion forceps, drawing it from its sheath and twisting till the inner and middle coats give way. It is efficient for even the largest arteries, but takes more time than other methods.

Describe acupressure.

This consists in checking hemorrhage by compressing the

wounded vessel between an acupressure needle and the tissues. The methods of accomplishing this are by—

1. *Circumclusion.* A pin or needle is thrust through the tissues, *beneath* the artery, and brought out to the surface on the opposite side. If necessary a thread can be carried around the two extremities of the pin in the form of a figure-of-**8**. The hare-lip suture is really an application of circumclusion.

2. *Torsoclusion.* The pin transfixes the tissues parallel to the artery, is twisted till it lies at right angles to its former direction, is pushed directly across the artery, and plunges into the tissues on the opposite side.

3. *Retroclusion.* The needle is carried in and out, transfixing the tissues on one side of the artery and at right angles to its course. The point of the needle is then carried over the artery to the opposite side, is plunged directly downwards, is carried *under* the artery and its point makes its appearance on the side from which it originally started.

Describe forcipressure.

Forcipressure consists in seizing the end of the bleeding vessel in hæmostatic forceps, which are allowed to remain in place till either the end of the operation, or till the forceps are required in another place, when they should be *gently* removed. The artery is crushed; the middle and inner coats break as in ligation.

What drugs may be administered by the mouth for the arrest of hemorrhage?

Opium, ergot, ol. erigeron., acid. sulph. aromat.

What is primary hemorrhage?

Bleeding which occurs immediately, on the infliction of a wound.

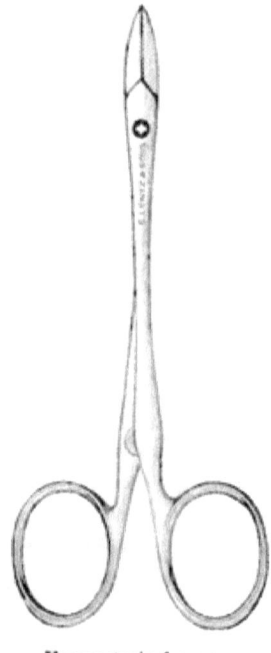

Fig. 1.

Hæmostatic forceps.

What is recurrent hemorrhage?

Synonyms: Reactionary, consecutive, intermediate.

Bleeding, which comes on with reaction. It occurs within the first twenty-four hours after a wound.

What are the causes of recurrent hemorrhage?

The *slipping of a ligature.* The *displacement of a clot.* This may occur from the wounded part not being kept at rest, or from the increased force of reaction circulation.

How do you treat recurrent hemorrhage?

First *elevate*, and apply *firm pressure* by means of additional bandages, covering in the soiled dressings with antiseptic gauze. If bleeding still continues, remove the dressing, open the wound, clear out the clots, and ligate or secure the bleeding vessel.

What is secondary hemorrhage?

Bleeding which comes on between the end of the first day and the complete cicatrization of the wound. It is most frequent about the time of the separation of ligatures or sloughs.

What are the causes of secondary hemorrhage?

1. *Constitutional conditions* which interfere with organization, or are associated with an overacting heart. Instance, Bright's disease, diabetes, hæmophilia, traumatic delirium, septicæmia, pyæmia, and plethora.

2. *Disease of the arterial walls*, as found in atheroma, calcareous degeneration, syphilis, or tuberculosis.

3. *Septic condition of the wound.* The ulceration and sloughing may involve the arterial walls.

4. *Defect in the ligature or its application.* The ligature may soften prematurely. It may be septic and cause suppuration. It may be badly applied, being too loose, or irregularly knotted, or tied too near a collateral branch.

How do you treat secondary hemorrhage?

If from a severed artery, as in a stump, and only a few days have elapsed since the infliction of the wound, treat as consecutive hemorrhage; that is, try elevation and pressure first, if the bleeding be moderate, that failing, or at once, in case of violent hemorrhage, reopen the wound and secure the vessel. If there is much sloughing use the actual cautery.

Later, when the healing is well advanced, try *pressure* first, then either reopen the wound, or ligate the main artery just above. If the bleeding recurs amputate higher up.

If from an artery tied in its continuity. *Pressure* by graduated compresses and compression of the artery above. If this fails *open the wound* and *tie above and below*. Should the bleeding still persist *amputate*, if the femoral artery is the one involved, or tie above, in the case of other arteries.

How do you cleanse wounds?

Gross foreign particles can be picked out with forceps. Blood clots and dust should be washed away by means of a fine stream of sterile or antiseptic liquid; avoid all rough handling or rubbing.

How do you provide for drainage?

By means of drainage tubes, which may be made of red rubber, glass, or decalcified bone; or by strands of catgut or horsehair. Drainage does not allow the serous exudate to make tension in the wound, or to remain as a rich culture fluid for the reception of germs. It should be employed in all wounds except those which are superficial, or are placed in very vascular regions, as in the face. Drainage tubes are to be removed in from 24 to 48 hours. If the wound is very deep and extensive take the tubes out gradually. The tube should be carried *through* the protective, should be cut off flush with the surface, and should be prevented from slipping into the wound by silver wire or a safety pin.

How do you close wounds?

Both *edges* and *surfaces* must be approximated. In superficial wounds adhesive plaster, isinglass plaster, or gauze collodion and iodoform may be used. In deep wounds *sutures* must be employed together with *compresses and bandages*.

Of what materials are the ordinary sutures made?

Silk, silver wire, catgut, horsehair.

Describe the various kinds of suturing.

1. *The continuous (glover's).* The stitches are made with one unbroken thread, carried across the wound in one direction.

2. *The interrupted.* Each stitch is carried across the wound and tied as inserted.

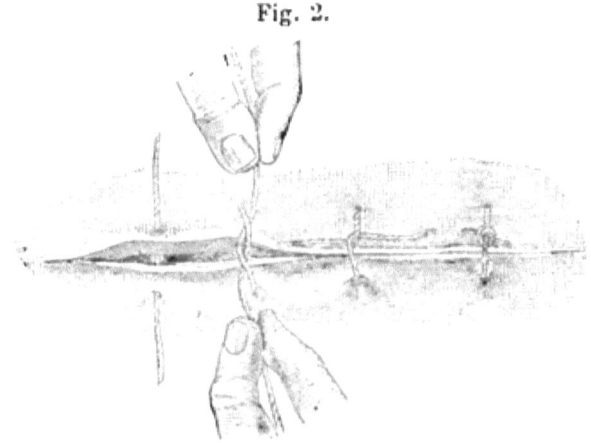

Interrupted sutures.

3. *Pin suture* (twisted or hare-lip). The apposed margins of a wound are transfixed with pins, around the two ends of which and across the wound is carried a thread in the form of a figure-of-eight. This keeps the surfaces in accurate apposition, and checks bleeding (circumclusion).

4. *The quill suture.* Threads are passed deeply across the wound and looped around quills or sections of catheter, placed parallel to the wound and at some little distance from its edges.

The button or plate suture. Wire is passed across the very bottom of the wound, brought out to the surface at some distance from its edges, and secured by fastening to leaden plates or buttons.

The Lembert and Czerny sutures will be described under intestinal wounds.

When there is much gaping, or loss of substance, the plate or quill sutures are used, they prevent tension in the skin sutures, and are termed *sutures of relaxation*. If the wound is moderately deep, a number of interrupted sutures are passed across

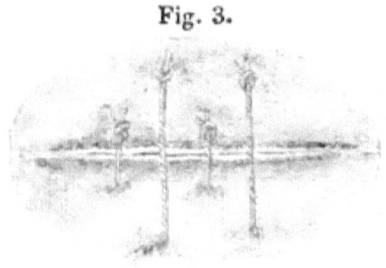

Fig. 3.

Sutures of approximation and coaptation.

it to its bottom and brought out at some little distance from its edges, these are termed *sutures of approximation*. The skin is accurately joined by closely applied superficial sutures, either interrupted or continuous, called *sutures of coaptation*.

Unless there is great tension, and reason to fear gaping, remove sutures about the fourth day.

How do you prevent putrefactive or infective processes in the wound?

By antiseptic treatment and dressing.

Describe the antiseptic treatment.

There must be provided *basins* for the sponges. *Shallow trays* for the instruments. *A fountain syringe* for irrigation.

Solutions. Carbolic acid 1 : 20. Bichloride of mercury 1 : 500. These solutions can be weakened by the addition of water as required.

Sponges and *drainage tubes* which have been kept in carbolic acid 1 : 30.

Ligatures and sutures which have been rendered aseptic and are kept in absolute alcohol.

The surgeon prepares himself by scrubbing his arms, hands, and *nails* with a brush, sublimate soap, and hot water; puts on his antiseptic coat and again washes his hands in sublimate solution 1 : 500.

The patient is prepared by a general hot soap bath, if possible. The entire region of the wound of operation is scrubbed with hot water and sublimate soap, shaved, and irrigated with 1 : 500 sublimate solution.

All portions of the patient's body and the operating table near the seat of injury are covered with towels wet in 1 : 500 sublimate solution. Instruments and drainage tubes are placed in 1 : 30 carbolic solution. The sponges are put in a basin and covered with bichloride, of the strength used for irrigating. The fountain syringe is filled with bichloride 1 : 2000. The dressings are cut to the proper size, and wrapped in bichloride towels.

During the operation or manipulation, irrigate occasionally with the bichloride solution, finally flushing out, if the wound be large, with a weak solution, sterile water or salt solution. Carefully guard against instruments, sponges, or hands coming in contact with non-sterilized surfaces.

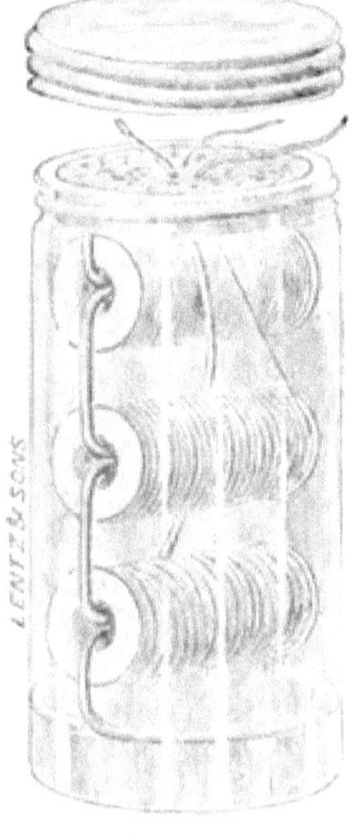

At the termination of the operation, see that the *hemorrhage is absolutely stopped*, and that *drainage is amply provided for*. Apply the dressing.

Fig. 4.

Sutures.

Describe the antiseptic dressing.

Lister's dressing. Dust with iodoform. Apply a piece of protective (varnished silk), wet in 1 : 40 carbolic, just large enough to cover the closed wound. Over the protective, and overlapping it, place several layers of carbolized gauze, wrung out in the 1 : 40 solution. Over this deep dressing and overlapping it, apply six layers of dry carbolized gauze, a seventh of Mackintosh (rubber cloth), an eighth of gauze. Over the whole and about the edges place antiseptic cotton, and cover in with a carbolized gauze bandage. *The protective* guards the wound surfaces from the irritation of the strongly carbolized gauze. The deep wet dressing disinfects the immediate neighborhood of the wound; it is wet because dry cold gauze may contain septic particles of dust. The Mackintosh prevents the discharge from passing through the gauze immediately to the surface.

The dressing in ordinary use is: 1. Dry iodoform gauze to the wound. 2. Covered and overlapped by bichloride gauze.

3. Bichloride cotton overlapping the whole and covered in by a gauze bandage.

When do you change an antiseptic dressing?

1. When drainage tubes or non-absorbable sutures are to be removed.
2. When fever, other than that due to reaction, appears.
3. When there is hemorrhage.
4. When the wound is healed.

Wounds.

What is a wound?

A solution in the continuity of the tissues, produced by sudden force.

Under what two headings may wounds be classed?

1. *Subcutaneous wounds.* There is either *no break* in the skin or an *exceedingly small one* compared to the extent of the lesion beneath. Instance, the wound of tenotomy is said to be subcutaneous.
2. *Open wounds.* The break in the surface is, to a certain extent, commensurate to the deeper injury.

What is a contusion?

A *subcutaneous* injury (distinguish from *contused wound* in which *there is a break in the surface*) occasioned by squeezing or crushing the tissues. There is hemorrhage and discoloration, at times vesicles and blebs form, and the part may appear gangrenous. The effused blood may form a fluctuating swelling, known as *hæmatoma*, or may coagulate, forming a hard swelling, termed *thrombus*.

How do you treat contusion?

By rest, pressure, and the application of evaporating and stimulating lotions.

Name the different kinds of open wounds.

1. Incised, or clean cut. 2. Lacerated, or torn. 3. Contused,

WOUNDS.

or bruised. 4. Punctured, or pierced. 5. Gunshot, or lacerated and contused. 6. Poisoned.

Describe incised wounds.

Cause. Sharp cutting instruments. They *bleed freely, gape widely,* and cause *burning pain.*

Treatment. Use all antiseptic precautions. Check hemorrhage by cold, forcipressure, and ligation. Bring the surface and edges of the wound in most accurate apposition. If tendons, nerves, muscles, or bones are severed, their corresponding ends must be carefully united by catgut sutures. If the wound is extensive, catgut drains may be employed. Absolute rest must be enforced. Union, in seven to ten days, by first intention

Describe lacerated and contused wounds.

Caused by machinery, dog-bites, blows with blunt instrument, etc.

Characterized by slight hemorrhage, moderate gaping, dull pain, ecchymosis (hemorrhage into the surrounding tissue), and shock.

Treatment. Antiseptic. Thoroughly cleanse, remove dead tissue, *provide for free drainage,* making counter openings in dependent positions, and using full-sized rubber drainage-tubes. Carefully coapt, if it can be done *without tension.* Apply iodoform gauze liberally, bichloride gauze, bichloride cotton, and bandages. Keep the part *absolutely at rest.*

Dangerous complications. Shock, extensive inflammation and sloughing, secondary hemorrhage, cellulitis, gangrene, tetanus.

Describe punctured wounds.

Caused by pointed instruments; depth is their greatest measurement. Usually associated with contusion.

Dangers. Wounds of deep structures, hemorrhage, the carrying in of septic substances, retention of discharge.

Treatment. Remove the vulnerating body, check bleeding, thoroughly disinfect the accessible portion of the wound, put in a drainage-tube, apply an antiseptic dressing, and put the part at rest. On the first sign of inflammation (pain and fever) remove the dressings, and lay the wound open to its very bottom; disinfect, drain, and reapply the antiseptic dressing.

Describe gunshot wounds.

Caused by missiles, either round (buck-shot, bird-shot) or conical (pistol and rifle balls). The wound of entrance is smaller than the wound of exit, and is slower in healing. One bullet may cause multiple wounds, depending upon the position of the wounded man and the direction from which the missile comes. Two bullets may form but one wound of entrance. One bullet may form several wounds of exit by being split in the body; the wound of entrance may also be the wound of exit, as, when a ball passes completely around the head, beneath the skin.

Balls may be deflected by tendons, bones, or even bloodvessels. Devitalization of tissue is proportionate to the velocity of the ball; hence is greatest at the wound of entrance.

The immediate effect of gunshot wounds is *hemorrhage, pain*, and *shock*. There may be no pain; excessive hemorrhage occurs only when large vessels have been wounded; shock may be delayed.

The secondary effect of gunshot wounds is inflammation, sloughing, hemorrhage, with the complications incident to contused and lacerated wounds (tetanus, gangrene, cellulitis, and blood poison).

How do you treat gunshot wounds?

On the field. Check hemorrhage by position, pressure, or the tourniquet. *Apply an antiseptic pad* to the surface wounds. Immobilize. If no septic matter has been carried in by the missile, or *the surgeon's probe or finger*, the wound is practically rendered subcutaneous by this treatment, and can be allowed to heal as such, no effort being made to find the ball.

In the hospital. Under all antiseptic precautions, remove the antiseptic pad, thoroughly clean the opening of the wound and the skin surface about it. Reapply an antiseptic dressing and immobilize. *Do not probe.* If inflammatory fever appears, or if the original wound was so extensive as to preclude the idea of primary occlusion, *do a formal antiseptic operation*. Freely lay open the wound tract, remove foreign bodies, devitalized tissues, or loose fragments of bone, explore and irrigate every recess of the wound, pack with iodoform gauze, insert sutures for the purpose of approximating the parts, but *do not tie them*, dress antiseptically. In one or two days remove the dressing and iodoform packing. If the wound is aseptic, close by knotting the sutures. If the wound is not aseptic, irrigate and renew the packing, or supply free drainage, dressing daily till the granulations become healthy.

An aseptic bullet is readily encysted. Should it subsequently give trouble, its removal is much safer after the wound has healed. If the surgeon decides to search for the bullet and extract it, he must proceed as in a formal operation.

Nélaton's probe, tipped with unglazed porcelain which is marked by contact with lead, and long-bladed *bullet forceps*, may be useful in locating and extracting a bullet.

What gunshot wounds require amputation?

1. Wounds which comminute the bone and injure or destroy the main vessels of a limb.
2. Wounds which destroy a large portion of the limb, or carry away a part of it.
3. Wounds complicated by osteomyelitis, intractable secondary hemorrhage, or spreading gangrene.

What injuries are classed as poisoned wounds?

1. Dissecting wounds. 2. Stings of insects. 3. Wounds inflicted by *arachnids* and reptiles. 4. Wounds infected from diseased animals.

Describe the dissecting wound.

It appears more frequently where fresh bodies or arsenical injections are dissected. It is due to inoculation with infective micro-organisms; these are destroyed by advanced putrefaction, hence the most offensive bodies may be the least dangerous. Its virulence depends upon the strength of the original virus and the constitutional vigor of the patient infected.

Symptoms. Within twenty-four hours of the infliction of a

scratch or cut, there is an itching, then a burning pain; a vesicle is formed which breaks, disclosing an indurated ulcer. There may be a stop at this stage, or the inflammation may extend; the lymphatic vessel and axillary glands become involved, and may suppurate freely. The constitutional symptoms are well marked. The patient may reach this stage and rapidly recover, or the disease may make steady progress, suppuration attacking the neck and thorax, cellulitis involving the arm, the symptoms becoming markedly adynamic, and the patient perishing of septicæmia or pyæmia.

How do you treat dissecting wounds?

Immediately, at the time of infliction, encourage bleeding by tying a ligature about the part. Suck the wound and press the blood from it; apply carbolic acid or sulphate of zinc, dust with iodoform, and cover with a light antiseptic dressing.

If an infective inflammation appears, freely incise, curette the indurated tissue, pack with iodoform gauze and dress antiseptically, applying a splint. Open abscesses *promptly*. Make multiple incisions for cellulitis.

Clear the bowels, give stimulants, tonics, and nutritious diet.

For pain, apply locally, chloral gr. xx. to the ounce of water.

A circular blister about the arm may limit the extension of lymphangitis.

There is always marked constitutional involvement in these wounds. There is fever and exhaustion, loss of sleep from pain, and the rapid development of an adynamic condition. Treat by anodynes, stimulants, full diet, tonics.

(For Anthrax, Glanders, Hydrophobia, *see* pp. 54, 55.)

How do you treat stings of insects and spider bites?

Locally. Ammonia.

Systemically. Stimulants if necessary, ammonia or brandy.

What are the symptoms of rattlesnake poisoning?

Rapid and extensive swelling, discoloration, and disintegration. Profound systemic depression.

How do you treat rattlesnake bites?

1. Put a *tight* ligature about the part above the wound.

2. Excise, and subsequently cauterize the wound area.
3. Encourage bleeding by suction.
4. Administer alcohol to the point of intoxication.
5. Release the ligature for a few seconds at a time, tightening again till each small dose of poison thus admitted to the system is eliminated. This is termed the *intermittent ligature*.

Injections of permanganate of potassium in and about the wound (10 per cent.) are said to be efficient. If collapse threatens, ammonia must be given hypodermically.

Wounds of Arteries.

Describe wounds of the arteries.

1. Non-penetrating. The outer coat or coats only are involved. The artery may subsequently ulcerate and give way, causing extravasation, or may cicatrize and gradually yield, forming *true circumscribed traumatic aneurism*.

2. *Penetrating*. The artery is laid open. It may be *partially* cut across, when there will be free and continuous bleeding, or *completely* cut across, when contraction and retraction favor coagulation.

How do you treat wounded arteries?

Ligation in the case of large and accessible arteries; forcipressure, acupressure, or the actual cautery under other circumstances. When the artery is partially divided, complete the division.

What rules must be observed in applying the ligature to a wounded artery?

Tie in the wound. Tie both ends of the wounded vessel. Do not search for the arterial wound unless there is actual bleeding at the time of search. While operating, check further bleeding by pressure, or by the finger in the wound.

How do you treat gangrene appearing after ligation of a wounded artery?

If rapidly progressive, amputate at once. If slow in progress, wait for the line of demarcation.

Describe traumatic aneurisms.

1. *Diffuse traumatic aneurism.* This is simply a collection of arterial blood, in the tissues of a part, which communicates with the blood stream in the interior of the artery, and is limited by peripheral coagulation.

2. *Circumscribed traumatic aneurism.* This is blood in the tissues, communicating with the arterial current, and provided with a sac formed by the condensation of the surrounding cellular tissues. The circumscribed traumatic aneurism may be formed by a protrusion of the inner coat through a laceration of the outer, in which case it is called *hernial;* or by the yielding of a cicatrix of the arterial coat, when it is called *true circumscribed traumatic aneurism.*

Symptoms as for aneurism, except in the case of *diffuse traumatic aneurism,* when a spreading tumor, in which thrill and bruit can be detected, and feeble or absent circulation of the part below, will indicate the nature of the affection.

How do you treat traumatic aneurism?

Ligate just above, or, if the aneurism threatens to burst, open the sac and tie above and below.

Describe an arterio-venous aneurism.

Definition. An abnormal communication between an artery and a vein.

Cause. A wound involving both vessels.

Varieties: 1. *Aneurismal varix.* The artery and vein communicate directly. The vein is dilated by the arterial beat, forming a fusiform swelling.

2. *Varicose aneurism.* The artery and vein communicate by means of an intermediate sac.

Symptoms. A tumor, characterized by a jarring pulse, and a rough buzzing bruit. The artery is large above and small below. The vein is large above and pulsates.

Treatment. Pressure on the tumor by means of an elastic bandage. Ligation of the artery above and below. When pressure fails to control the bleeding from the vein, it must be ligatured also.

What are the dangers in wounds of veins?

1. *Hemorrhage.* Control by pressure or ligation.
2. *Blood poisoning* from septic thrombosis. Prevent by keeping the wound aseptic.
3. *Entrance of air.* Characterized by a hissing sound during inspiration, by the escape of frothy blood during expiration, by a churning sound heard on ausculting the heart, and by prompt collapse of the patient. Stop the vein wound immediately with the finger, or fill the entire wound with water. Ether, brandy, or ammonia subcutaneously.

Wounds of Nerves.

What are the consequences of wounded nerves?

The nerve may be partially or completely divided. If completely divided, the entire peripheral part undergoes atrophy and degeneration (Wallerian degeneration), the proximate end becomes bulbous from proliferation of the fibrous tissue. Should union occur the degenerated fibres are regenerated.

As a result of destroyed innervation there follows :—

1. Motor and sensory paralysis.
2. Muscular atrophy and degeneration.
3. Trophic changes, characterized by the skin becoming glazed, smooth, bluish-red, and prone to ulcerate; the nails becoming cracked and deformed; the hair falling out; and rheumatoid joint affection.

How do you treat wounded nerves?

If recent, suture together with fine chromicized catgut passed through the sheath of the nerve. If old, free from all cicatricial adhesions, resect the bulbous proximal extremity, freshen the distal extremity, and suture as before.

Head Injuries.

Give the surgical anatomy of the scalp.

Layers. Skin, superficial fascia, aponeurosis of the occipito-frontalis, subaponeurotic fascia, pericranium.

Superficial fascia binds the skin firmly to the aponeurosis. It is made up of intersecting, non-elastic bands of connective tissue, containing in its meshes globules of fat; it is very vascular, and freely supplied with nerves.

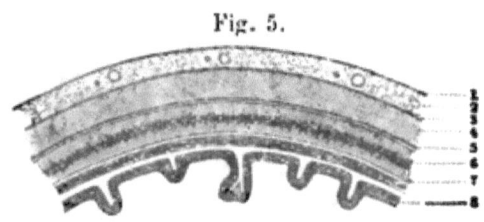

Layers of the scalp.

Aponeurosis. Covers the vault of the skull, is attached to the superior curved line and the mastoid process; is blended in front with the pyramidalis nasi, corrugator supercilii, and orbicularis palpebrarum, and is continued laterally to the zygoma by laminated layers of areolar tissue.

Subaponeurotic fascia. Is made of delicate, elastic, connective-tissue fibres containing no fat; loose in texture, and allowing free motion on the part of the aponeurosis. Blood supply limited.

Arteries of the scalp are from the temporal, occipital, auricular, supraorbital, and frontal. Certain branches strike deep and supply the periosteum.

Veins of the scalp intercommunicate with those of the pericranium, the diplöe, the meninges, the sinuses.

What is the surgical bearing of these facts?

1. From the vascularity of the superficial fascia extensive injury can be quickly repaired.

2. From its lack of elasticity no tension can be made in uniting wounds. There is little gaping unless the aponeurosis is cut.

3. From its denseness of structure, effusion, or suppuration will probably be circumscribed, and movable only to the extent that the aponeurosis can be moved.

4. In the subaponeurotic fascia effusion or suppuration will

probably not be circumscribed, from the looseness of structure, and will appear as a fluctuating swelling about the ears or the root of the nose, from which position it can be moved to the various dependent parts of the aponeurotic attachment.

5. The arrangement of the vessels allows the scalp to be entirely detached from the pericranium without loss of vitality.

6. It also allows of the direct extension of septic processes into the diploë and the interior of the skull.

7. Swellings beneath the pericranium are bounded by the sutures and are immovable.

Describe contusion of the scalp.

Swelling very rapid. On palpation a *soft yielding centre* (fluid blood), and hard, *distinctly outlined edges* (fat and coagulum).

How do you diagnose contusion from depressed fracture?

The hard margins about the apparently depressed central area are *raised from the bone*. By firm pressure with the nail the clot may be pushed aside, and the bone felt through it.

In case of fracture, the finger passes *directly from the surface of the skull into a depression*, without first surmounting a ridge.

Where may the effusion due to contusion take place?

The blood may be effused in the superficial fascia, beneath the aponeurosis and beneath the pericranium. When in the latter position it may ossify.

How do you treat contusions of the scalp?

Ice-bag till swelling ceases to increase. Evaporating and stimulating lotions, moderate pressure. Aspirate a persistent hæmatoma. If suppuration occurs, incise freely.

How do you treat wounds of the scalp?

Carefully shave, wash, and disinfect the region of the wound. Remove all foreign matter, and check hemorrhage. If the wound is very extensive, drain by strands of horsehair or catgut. Suture, making accurate apposition, apply iodoform, protective, wet bichloride gauze, dry bichloride gauze, bichloride cotton, and a firm bandage.

Describe contusions of the cranial bones.

Contusions may cause—

1. An inflammation of the pericranium, or periostitis, which may terminate in *resolution, chronic periostitis, or suppuration,* involving the neighboring bone, and terminating in *caries* or *necrosis.*

2. The inflammation may extend to the diploë, causing septic osteophlebitis, with septicæmia or pyæmia.

3. The inflammation may extend to the intracranial structures, causing supra- or subdural suppuration.

4. The inflammation may terminate in chronic osteitis and pachymeningitis, causing thickening.

What symptoms aid the surgeon in determining the character and seat of inflammatory action?

1. *Pus beneath the pericranium, or simple necrosis.* Chill and fever, moderate in severity, local œdema, tenderness, and deep fluctuation. Detection of the diseased bone when the abscess is opened.

2. *Pus in the diploë.* Chill, high fever, local signs of suppuration, general symptoms of pyæmia or septicæmia.

Intracranial extension. High fever, headache, vomiting, monoplegia or hemiplegia, delirium or stupor.

Pott's puffy tumor, a circumscribed superficial swelling over the affected area, sometimes accompanies supradural suppuration.

How do you treat contusions of the cranial bones?

Open the bowels freely, keep the patient in bed and absolutely quiet, give liquid diet, and apply cold to the head. If there is a wound, rigid antisepsis must be observed. Should symptoms point to subpericranial suppuration, open freely. Deeper suppuration should at once be exposed by the trephine.

Classify fractures of the skull.

A. Fractures of the vault. B. Fractures of the base.

1. *Partial,* involving the inner or the outer table.

2. *Complete,* involving the entire thickness of the skull. The inner table is usually damaged more extensively than the outer.

Of the *complete* fractures we have—

1. *Fissured*, taking the form of a simple crack.
2. *Stellate* or *radiate*, appearing as several fissures radiating in different directions.
3. *Comminuted*. The bone is broken into several pieces.
4. *Depressed*. The bone is pressed in upon the brain.
5. *Punctured* or *pierced*. This is usually accompanied by considerable comminution of the inner table.

Any of these fractures may be *simple* (no external wound) or *compound* (external wound communicating with the break).

What causes fractures of the vault of the skull?

Sudden concentrated force, as the blow of a hammer.

How do you diagnose fractures of the vault of the skull?

Simple fractures without displacement (fissured, stellate) can only be inferred from accompanying symptoms.

Simple fractures with displacement can frequently, but not always, be detected by careful examination of the surface. There is usually *depression*, and the abrupt bone edges may be felt. Symptoms of compression are commonly present.

Compound fractures can be diagnosed by inspection and palpation through the wound. There is frequently free bleeding, and there may be escape of cerebrospinal fluid.

How do you treat fractures of the vault?

Simple or compound fracture, without depression.

Place the patient in a quiet, darkened room, clear the bowels with calomel, shave the head, and apply an ice-bag; give a light milk diet (Oij daily). If the wound is compound, treat antiseptically.

Calomel gr. $\frac{1}{6}$, Dover's powder gr. ij, every two hours, is sometimes kept up for three or four weeks.

Simple depressed fractures without signs of compression treat as above unless symptoms arise.

Compound depressed fractures, and punctured fractures. Always elevate, trephining if necessary. Thorough asepsis makes the operation entirely safe. Punctures through the supraorbital plate or the nose do not in themselves indicate trephining, though the operation should be done if unfavorable symptoms subsequently appear.

What is the cause of fractures at the base of the skull?

Direct force. Punctures. Driving of a condyle through the glenoid fossa by a blow upon the chin, or shattering the cribriform plate of the ethmoid by a blow on the nose.

Indirect force. 1. *Falls upon the buttocks* or feet drive the spine against the occipital condyles.

2. *Falls upon the cranial vault* drive the occipital condyles against the spine. If the head is flexed the force is carried backward, and is exerted on the posterior cerebral fossa. If the head is extended, the force is carried forward, and is exerted on the anterior or middle cerebral fossa.

3. *Conduction* and *amplification of vibrations.* The force is powerful and diffused. If applied to the *frontal region*, there is usually fracture of the anterior cerebral fossa. The *middle cerebral fossa* is fractured by such force applied to the *temporo-parietal* region. The *posterior cerebral fossa* by force applied to the occipital region.

What are the symptoms of fracture of the anterior cerebral fossa?

Free and continuous bleeding from the nose. Subconjunctival effusion with palpebral ecchymosis, involving the lower eyelid particularly. Escape of watery fluid (cerebro-spinal fluid) from the nose. Paralysis of the olfactory, optic, or oculo-motor nerves. Concussion or compression.

The blood and cerebro-spinal fluid may pass back into the pharynx, which should always be examined in these injuries.

What symptoms denote fracture of the middle cerebral fossa?

Free continued bleeding from the ear, followed by escape of cerebro-spinal fluid, increased in quantity by firm pressure on the jugular veins.

Paralysis of the auditory and facial nerves, usually coming on some days after the injury. If the membrana tympani is not ruptured, the blood and cerebro-spinal fluid will escape into the pharynx by way of the Eustachian tube.

What symptoms characterize fractures of the posterior cerebral fossa?

Examination through the pharynx may show depression or comminution. Severe pharyngeal hemorrhage. Ecchymosis of the lateral regions of the neck.

When the neck is not involved in the injury late discoloration is a valuable sign of fracture at the base (middle or posterior fossa).

How do you treat fractures of the base?

Since these fractures are usually fissured, they, in themselves, rarely require treatment. The gravity of fractures of the base depends almost entirely upon the concomitant injury to the brain or its bloodvessels, and the treatment must be directed to the prevention of encephalitis which is liable to develop after these injuries.

Keep the patient absolutely quiet. Elevate the head and apply an ice-bag to it. Control restlessness by bromide of potassium or morphia. Give water only, for 48 hours, then a light liquid diet. Mercurials may be used.

When the cerebro-spinal fluid escapes externally, the fracture is, of course, compound, and the channel of escape must, if possible, be antiseptically cleansed and occluded.

Injuries of the Meninges and Brain.

In what regions may intracranial blood extravasations take place?

1. Between the dura mater and the skull.
2. In the cavity of the arachnoid.
3. In the meshes of the pia mater (on the brain surface).
4. In the cerebral substance.
5. In the ventricles.

What are the sources of extravasation between the dura mater and the skull?

1. The small vessels passing from the dura to the bone. The hemorrhage is slight in amount.

2. *The middle meningeal artery.* The usual source of extensive bleeding.

3. The venous sinuses. Rarely a source of bleeding.

What symptoms denote extravasation of blood between the dura mater and the skull?

Symptoms of compression *coming on after an interval of immunity.*

Immediately after an injury the patient suffers from *concussion* and *shock;* he reacts and recovers from this condition shortly to exhibit symptoms of *compression*, characterized by: 1. Spasm followed by *paralysis*, affecting the face, arm, or one side of the body, and accompanied by a local fall of temperature. 2. Coma. 3. Widely dilated pupil of the affected side.

How do you treat hemorrhage between the dura and the skull?

Trephine over the middle meningeal artery (anterior branch). The pin of the trephine is placed 1½ inches behind the external angular process of the frontal bone, and the same distance above the most prominent part of the zygoma. Clear away the clot, close the artery by means of ligatures, a plug of wax or catgut, or the touch of a hot needle. If the trephine opening does not expose the bleeding point, remove the bone along the course of the artery till the source of hemorrhage is found.

If no supradural hemorrhage is found, but the dura is bluish, projecting, and does not pulsate, there is effusion beneath, which must be evacuated by incision.

If the symptoms do not definitely indicate the probable seat and nature of the injury, treat as for all head injuries, *i. e.*, elevate the head, and apply cold to it, clear the bowels, give a very restricted fluid diet, use bromides, chloral, morphia, mercury, or bleeding as indicated by symptoms.

What are the symptoms of hemorrhage beneath the dura?

Blood in the arachnoid is generally diffused over the whole cerebral hemisphere. There may be symptoms of compression, or, some time after the injury, irritability of temper, headache, or convulsions may develop. There is nothing diagnostic. The effused blood may become encysted or may organize as a tough membrane.

Blood in the pia mater usually accompanied by cerebral laceration. The blood is widely diffused. The symptoms are those of the brain injury, or of apoplexy.

How do you treat subdural extravasations?

Expectantly, as for head injuries in general. If the symptoms should point to localization of the hemorrhage, trephine.

Concussion and Contusion.

Describe concussion of the brain.

By concussion is meant a simple jarring of the brain without attendant lesions. There is, however, always congestion, and, commonly, serous or sanguinolent effusion. If concussion is attended with marked and persistent symptoms, it is probably associated with contusion.

Contusion may be circumscribed or diffused. It may produce hemorrhage in mass, or diffuse miliary extravasations. Its effects may be found at the point of injury, or on the opposite portion of the brain. Laceration frequently accompanies contusion. The anterior part of the frontal and temporo-sphenoidal lobes are commonly involved.

What are the symptoms of concussion?

Of the slighter form, momentary loss of consciousness, or giddiness, with pale face and feeble pulse, some mental confusion, sweating of the face, nausea, vomiting, and reaction.

Of the more severe forms (contusion, with congestion, bleeding or laceration), prolonged unconsciousness, with feeble, scarcely perceptible pulse, shallow breathing, pale, cold surface, subnormal temperature, muscular relaxation, variable pupils (dependent on the seat and character of the injury). Restlessness, screaming, and local spasm or paralysis may suggest laceration. The *beginning of reaction* is characterized by *vomiting*.

After a variable time the patient may pass into the *second stage of concussion*, termed *cerebral irritation*.

He can be roused with difficulty, but responds angrily, and immediately lapses into a somnolent condition.

He lies curled up on his side, with limbs flexed and eyes tightly closed. He resents any effort at changing his posture. He may be exceedingly restless.

The pulse is small and feeble, the respirations are quiet, or at least are not stertorous. The pupils are contracted.

As the condition of cerebro-irritation subsides, the *third stage of concussion*, characterized by inflammation, abscess, softening, or fatuity, may develop. Later, hereditary or acquired tendency to brain disease may appear.

Concussion and contusion are always attended by *shock*.

How do you treat cerebral concussion and contusion?

First stage (insensibility and shock). Absolute quiet in a darkened room. If reaction is slow, encourage by external heat. *Very rarely* should stimulants be given; if *absolutely indicated*, administer brandy or ammonia hypodermically. On the development of the *second stage* (cerebral irritation) apply an ice-bag to the raised head, clear the bowels, give water and cracked ice for two days, followed by milk and lime-water, in small quantities. For restlessness and pain give bromide, chloral, or opium. Prevent sequelæ by long-continued rest in bed, by very slow resumption of ordinary duties and responsibilities.

Compression.

What are the causes of cerebral compression?

1. Depressed bone. 2. Extravasated blood. 3. Pus, or inflammatory products. 4. Foreign bodies. 5. Tumors.

What are the symptoms of cerebral compression?

Unconsciousness, absolute (coma). *Respirations*, slow, stertorous, blowing. *Pulse* full and slow. *Paralysis* involving one side of the body. *Pupils* may be unequal. *Urine* retained, fæces passed involuntarily. *Decubitus* dorsal.

How do you determine as to the cause of compression?

Symptoms appear immediately when due to depressed fracture

or foreign body; after some hours, if due to hemorrhage; after some days, if due to inflammation.

How do you treat compression of the brain?

Trephine and remove the cause, if it can be located. Under other circumstances expectantly, as for head injuries in general.

How do you distinguish concussion from compression?

In many cases this cannot be done; the symptoms of one condition merging into those of the other. The distinctive symptoms of the two affections are as follows (Agnew):—

Concussion.	Compression.
Patient semi-conscious; special senses blunted, not abolished.	Absolutely unconscious, paralyzed, and with abolition of special senses.
Power of movement not lost.	
Respiration quiet and feeble.	Respiration full and noisy.
Pulse feeble, frequent, and intermittent.	Pulse full, slow, laboring.
Nausea and vomiting.	Neither nausea nor vomiting.
Pupils generally contracted.	Pupils generally dilated, often unequal.
Subnormal temperature.	Temperature about normal.

Of what significance is the size of the pupil in brain injuries?

A *contracted pupil* denotes cerebral irritation (slight injuries or effusion). A pupil *fixed in wide dilatation* denotes abolition of cerebral function (large effusions or extensive injury).

Intracranial Inflammation.

What are the causes of traumatic intracranial inflammation?

Wounds of the scalp, bone, or brain. *Fractures* or *contusions* of the cranial bones. *Concussion, compression, contusion,* or *laceration* of the brain.

Describe traumatic intracranial inflammation.

There may be either *meningitis* or *encephalitis*. More commonly, both meninges and brain are involved (meningo-encephalitis). Should suppuration occur, the pus may be *diffused*, or may form an *abscess*. The inflammation may be acute or chronic.

Give the symptoms of traumatic intracranial inflammation.

Pain referred to the seat of injury, *fever, intolerance* of light and sound, *vomiting* with a clear tongue, *contracted pupils, quick, full pulse*, restlessness, insomnia, and delirium. Later, compression symptoms develop, and the patient perishes comatose. Formation of pus is attended by *rigors*.

How can you localize the inflammation?

If, in from one to four weeks from the infliction of injury, symptoms of encephalitis suddenly develop preceded by headache, if Pott's puffy tumor of the scalp forms, if there is local spasm or paralysis, and the history of a chill, there is probably an abscess between the dura and the skull.

Inflammatory symptoms, appearing about the fourth day after a head injury, point to contusion or laceration of the brain substance.

If, after several weeks, there is found optic neuritis, with hebetude, headache, and involvement of motor areas; if there has been a chill, and symptoms of compression develop suddenly, there is probably a cerebral abscess.

How do you treat traumatic meningo-encephalitis?

Prevent by quiet, cold to the head, purgation, low diet, and *absolute asepticity* of all head wound.

Treat, on the earliest symptom, by calomel, bleeding from external jugular, ice-bag to head, light diet; opium and bromide as required, calomel gr. $\frac{1}{3}$. Dover's powder gr. ij every two hours.

If an abscess can be localized, trephine and evacuate.

Describe hernia cerebri.

Definition. A protrusion of brain matter disintegrated by inflammatory action, through an opening in the skull.

Cause. Wound of the bone and dura mater, attended with laceration and bruising of the brain substance.

Appearance. A blood-stained, fungous mass, projecting from the skull opening.

Prognosis. Usually bad.

WOUNDS.

Treatment. Remove all irritating causes, such as spiculæ of bone. Treat in general as for encephalitis.

Locally, apply antiseptic dressings, with *very moderate* compression. Nature sometimes effects a cure by strangulating the growth.

What are the prognosis and treatment of foreign bodies in the brain?

The *ultimate* prognosis is bad in all cases where the foreign body is not *removed*. The usual foreign body is a bullet. Its wound may be *perforating* or *penetrating*.

The *perforating wound* allows of free drainage, and the foreign body has passed out; hence, if not intrinsically fatal, the prognosis is comparatively favorable. Trephine, if necessary.

The *penetrating wound* should be trephined to remove bone spiculæ. Explore with a soft rubber catheter. The ball, being found, should be removed, either through the wound of entrance, or by making a counter trephine opening. Provide abundantly for drainage. *Absolute asepsis.* Treat as for head injuries.

Cerebral Localization.

Give the position of the motor areas grouped about the fissure of Rolando.

1. *The face.* Motor and sensory nerves from lower third of the ascending frontal and parietal convolutions, and posterior end of the second frontal convolutions.

2. *The arm.* Motor and sensory supply from middle third of ascending frontal and parietal convolutions.

3. *The leg.* Motor and sensory supply from the upper portion of the ascending frontal and parietal convolutions, and the paracentral lobule.

4. *The tongue.* Receives its nerve supply from the posterior portion of the third (inferior frontal) convolution of the left side in right-handed persons.

Local spasm and hyperæsthesia indicate an irritative lesion of a motor area.

Local paralysis and anæsthesia indicate complete suppression of function from more extensive injury.

What symptoms founded on cerebral localization indicate trephining?

Hemiplegia, complete or incomplete, with or without hemispasm, following a blow on the temporo-parietal region, would indicate an exploratory operation on the side opposite to that of peripheral symptoms.

Monoplegia or monospasm following an injury to the head indicates operation.

Mono-hyperæsthesia—anæsthesia or analgesia following an injury indicates an operation.

If the peripheral sensory or motor disturbance be on the side opposite to that of the lesion, operate at the site of the lesion; if, however, these symptoms are on the same side, exploratory operation would be indicated on the opposite side of the head.

What symptoms contraindicate operation?

Lesions of the base of the brain as indicated by paralysis of cranial nerves, neuro-retinitis, Cheyne-Stokes respiration.

Hemiplegia accompanied by anæsthesia.

How can the position of the Rolandic fissure be indicated upon the head?

Shave the scalp, draw a vertical line from one external auditory meatus to the other (at right angles to the alveolo-condyloid plane), from the centre of this vertical line (bregma) measure directly backward for 5.5 centimetres (5 in women). From the external angular process of the frontal bone measure 7 centimetres horizontally backward and 3 centimetres vertically upward: a line drawn from this point to the point 5.5 centimetres posterior to the bregma will indicate the fissure of Rolando. For general hemiplegia trephine over the centre of the line. In other cases over the portion chiefly involved.

Sensory disturbances of the arm or leg would indicate that the lesion lies somewhat *posterior* to the fissure of Rolando.

What are the indications for trephining?

1. Simple depressed fractures, attended with persistent grave symptoms.
2. Compound depressed fractures. *Except* in children, when the depression is of less serious consequence and often spontaneously corrected.
3. Punctured fractures.
4. The presence of a foreign body.
5. Traumatic osteomyelitis and necrosis.
6. Localized blood clot between the dura mater and the bone.
7. Localized intracranial suppuration, with symptoms of compression or irritation.
8. Traumatic epilepsy or localized obstinate headache following an injury.
9. Accessible cerebral tumors.

Many surgeons advise trephining in all *depressed fractures*, with or without serious symptoms.

Describe the operation of trephining.

Prepare the patient the day before the operation, if possible, by shaving the scalp and washing with sublimate soap and warm water, followed by a cleansing with ether, after which washings with the sublimate soap and water must again be repeated. Apply, for twenty-four hours, to the entire scalp, gauze saturated in 1 : 2000 bichloride solution, covered in with an antiseptic dressing. Renew the sublimate and ether washings just before the operation, and further cleanse the surface with 1 : 500 bichloride solution.

The instruments required are scalpel, hæmostatic forceps, periosteal elevators, a conical trephine, a fine probe, a small stiff brush, a Hey's saw, bone forceps, curved needles, and catgut.

The incision. Must be *free* and to the *bone*, including periosteum. A semicircular flap is raised, the pin of the trephine is pressed to the bone, and, by a twisting motion, made to penetrate till the teeth grip, when the pin is withdrawn, and the instrument steadily worked through. Free bleeding indicates when the diploë is reached. (Note that in infancy and old age there is practically no diploë.) The instrument must now be advanced with the greatest care. It is removed from time to time, and the groove probed to see whether the inner tablet is penetrated at any part. When the bone is loosened, it is removed by means of sequestrum forceps or an elevator, and wrapped in a warm antiseptic towel. The surgeon now endeavors to accomplish the specific object for which the skull was opened. Spiculæ of bone are removed, depressed fractures are elevated, bleeding meningeal arteries are secured by passing a thread beneath them, clots are cleared away. If further exposure is necessary, it can be accomplished by dividing the bone by a chisel, bone forceps, or, best of all, a circular saw run by a surgical engine. On the completion of the operation free drainage is provided for by means of catgut strands, the disk of bone is replaced, either entire or cut into pieces, the flap is held in place by one or two sutures. Iodoform is dusted over the line of incision, a deep dressing of iodoform gauze is applied over and about the wound, and the dressing completed by bichloride gauze, bichloride cotton, and an elastic bichloride bandage.

Wounds of the Face.

What rules should be observed in treating wounds of the face?

Secure most accurate coaptation. Avoid sutures in superficial wounds, closing by means of iodoform, ether, and collodion. In wounds involving the cartilages of the nose or ear, pass sutures only through the skin. In operations, so place the incision that it may correspond with the natural lines of the face. If stitches are inserted, remove them in twenty-four hours.

How do you treat salivary fistula?

This is usually caused by a wound of Steno's duct. Treat by passing a thread around the duct from the *inside* of the cheek *posterior* to the external opening. When this thread has ulcerated an opening into the mouth, the external wound will usually heal. If not, freshen its edges and suture.

Wounds of the Neck.

(For the anatomy of the Cervical Triangles, see Ligations.)

Describe wounds of the neck.

These wounds are commonly incised suicidal wounds. They extend *obliquely* from left to right, and from above downward, and are *deepest* at their starting-point. They are most frequently found in the laryngeal region, particularly over or through the *thyrohyoid membrane*. The carotid arteries are rarely injured, the wound being usually placed too high, and the larynx and trachea bearing the brunt of the incision. These wounds may be *penetrating* or *non-penetrating*.

Wounds above the hyoid bone may divide the tongue, the lingual and facial arteries, and the hypoglossal nerve. There is great gaping; frequently escape of food and saliva.

Wounds through the thyro-hyoid membrane open the pharynx, and may involve the epiglottis, the superior thyroid and lingual arteries, and the superior laryngeal nerves.

Wounds through the cartilages may involve the vocal cords and the recurrent laryngeal nerve. There is usually but moderate bleeding.

Wounds below the cartilages may involve the superior or inferior thyroid arteries, the thyroid and anterior jugular veins, the trachea, and even the œsophagus.

What are the immediate dangers of penetrating neck wounds?

1. Hemorrhage, arterial or venous.
2. *Suffocation* from the plugging of the air-passages, with either blood-clot, the tongue, the epiglottis, or the divided cartilages.
3. Entrance of air into the veins.

What are the secondary dangers of penetrating neck wounds?

Œdema of larynx, emphysema, bronchitis or broncho-pneumonia, cellulitis, cicatricial contraction and stricture.

How do you treat penetrating neck wounds?

Check bleeding, ligate both ends of every bleeding vessel. The common carotid should only be tied for bleeding from its branches, when it is found impossible to tie the branches. If the external carotid is wounded at its origin, tie the common carotid, the external carotid, and, to avoid bleeding from collateral circulation, the internal carotid.

If the larynx is obstructed by blood-clot, clear by the fingers, by suction, or by forcing the air suddenly from the chest. Remove a partially severed portion of the epiglottis. Hold the divided tongue forward by a ligature passed through its tip.

Wounds of the œsophagus should be closed by catgut sutures. If the trachea is completely divided across, the two ends may be held in apposition by fine catgut sutures passed through the investing cellular tissue. The external wound should not be sutured; its surfaces are apposed by raising the head, and supporting it in one position by pillows and sand-bags, or by a gutta-percha splint.

Provision is made for free drainage, and light antiseptic dressing is applied. If dyspnœa appears, perform tracheotomy lower down, or insert a tracheal canula through the wound. Feed by the rectum for four days, then by an œsophageal tube, passed just beyond the wound. Non-penetrating wounds are treated as wounds in any other part of the body.

Wounds of the Chest.

Describe non-penetrating wounds of the chest.

A non-penetrating chest wound is one which does not involve the costal pleura. In chest wounds *the finger* must be used as a probe, and great care taken lest a non-penetrating be converted into a penetrating wound. Hemorrhage must be *absolutely checked* before closing, and the wound approximated by deep sutures passed to its very bottom. Firm pressure is applied over the antiseptic dressing, by a bandage carried around the chest.

These wounds may involve the brachial plexus, the intercostal, internal mammary, acromio-thoracic, long thoracic, or axillary arteries. Check bleeding by ligature or hæmostatic forceps.

Describe penetrating wounds of the chest.

The pleura and lung, the pericardium and heart, or the great vessels may be wounded.

Injuries of the pleura and lung are characterized by *shock*, dyspnœa, pain, cough, abdominal breathing, *expectoration of frothy blood-stained mucus, escape through the wound of a bloody froth accompanied by a hissing sound* (*traumatopnœa*), emphysema, pneumothorax, external bleeding, hæmothorax. In case the pleura alone is injured there will be no hæmoptysis and no bloody froth from the wound.

Prognosis, grave in wounds involving the root of the lung, and in gunshot wounds which penetrate but do not perforate.

Injuries to the pericardium and heart are characterized by great shock, hemorrhage, and the subsequent development, if the patient lives long enough, of pericarditis. Death in wounds of the pericardium occurs from shock, the pressure effect of hæmopericardium, or from pericarditis.

What are the complications of penetrating wounds of the chest?

External bleeding, hæmothorax, emphysema, pneumothorax, pleurisy, pneumonia, prolapse of lung.

How do you treat the external bleeding of penetrating chest wounds?

If from an intercostal artery ligate, or apply hæmostatic forceps; this being impossible, dissect off the periosteum from the lower part of the rib (carrying the artery with it of course) and tie; or resect a portion of the rib. A ligature may be carried around the entire rib.

If from the internal mammary, ligate in the wound, resecting the chondral cartilages if necessary.

If from the lung, close the external wound, place the patient on the injured side, and apply an ice-bag. Internally give opium, ergot, gallic acid. If the bleeding continues, producing constitutional signs of hemorrhage, and local signs of extensive hæmothorax, open again and allow the blood to escape.

Describe hæmothorax.

Definition. Bleeding into the pleural sac.

Usual cause. Wound of the lung, or of an intercostal artery by a broken rib.

Symptoms. Those of internal hemorrhage, together with bulging of the intercostal spaces, increasing dyspnœa, flatness on percussion, and absence of breathing sounds. The symptoms appear *almost immediately* after the injury. Inflammatory effusions do not take place till some days later.

Treatment. As for external bleeding from lungs. Aspirate or open if there is threatening dyspnœa. If suppuration takes place, open freely and drain.

Describe pneumothorax.

Cause. Injury to lung and pleura, usually by a broken rib.

Symptoms. The lung collapses. Increasing dyspnœa, great percussion resonance, amphoric breathing, metallic tinkling, bulging of intercostal spaces.

Treatment. Should dyspnœa become urgent, aspirate.

Describe emphysema.

Cause. Wound of the lung and pleura. It may arise after wound of the lung alone, in this case extending by way of the root to the posterior mediastinum, and from there into the connective tissue of the neck and arms.

Symptoms. A diffused, colorless, elastic, puffy swelling, crackling on pressure.

Treatment. A compress and bandage over the wound. Should distension become great, puncture.

How do you treat prolapse of the lung?

Return if not adherent. If adhesions have taken place, ligate or excise, taking precautions against opening the pleural cavity.

Describe hernia of the lung.

Causes. The yielding of a cicatrix. The result of subcutaneous wound. Great muscular effort.

Symptoms. A soft circumscribed tumor, resonant on percussion, giving a loud respiratory murmur, and crepitating on manipulation.

Treatment. Protective.

What is concussion of the lung?

A condition following traumatism. Characterized by dyspnœa, feeble respiratory murmur, and slight dullness on percussion. The symptoms pass off after a few hours.

What operations may be done for the evacuation of blood or inflammatory effusion within the chest walls?

1. *Tapping the pleura.* For serous effusion. Thrust an aspirating needle through the sixth intercostal space, in the mid axillary line. This operation must be done under antiseptic precautions. The skin is drawn down before the puncture is made, forming a valvular wound. Dress with iodoform and collodion.

2. *Incision and drainage of pleura.* For empyema and the removal of decomposing clots. Operate in the sixth intercostal space, in the axillary line, or as low as the eleventh space, in a line with the angle of the scapula. Make a careful dissection. Excise a portion of the rib if necessary, and insert a drainage tube.

3. *Tapping the pericardium.* Fourth intercostal space two inches to the left of the sternum.

4. *Incision and drainage of pericardium.* Beginning one inch from sternum, make an incision two inches in length along the upper border of the fifth or sixth ribs. Dissect down carefully, insert drainage tube after opening.

5. *Pneumotomy.* Lung incision for abscess, gangrene, or cysts. Open down to the pleura, thrust a trocar and canula into the affected area. Enlarge this puncture by dressing forceps.

Wounds of the Abdomen.

Describe contusion of the abdomen.

Contusion may take place with, or without, rupture of the contained viscera.

Contusion without rupture of the contained viscera is characterized by pain, discoloration, swelling, and shock. The rectus muscle may be ruptured, or there may be a hæmatoma formed, followed by abscess.

Treatment. Put the patient to bed, apply heat to the body, hot fomentations to the abdomen. Give water and cracked ice for twenty-four hours. Treat rupture of the rectus by position. Apply cold in case of hæmatoma. Evacuate abscesses early.

What symptoms denote contusion with laceration of the viscera?

Great shock, pain, persistence of collapse with signs and symptoms of *internal bleeding*, in case the solid viscera or a highly vascular portion of the peritoneum is ruptured, symptoms of *rapidly developing peritonitis* in case the hollow viscera are ruptured.

The following signs, *if present*, are indicative of rupture of the individual viscera.

Liver. Pain in right hypochondrium, increased hepatic dullness, signs of internal bleeding; later, bilious vomiting, clay-colored stools, sugar in the urine.

Spleen. Pain in left side, increased splenic dullness.

Stomach. Intense pain in stomach, hæmatemesis, rapid development of general meteorism, tympany over the liver.

Intestines. Intense radiating pains. Vomiting of stomach contents, then bile, finally blood. Bloody stools. Tympanites with dullness in the flanks. Percussion resonance over liver. Peritonitis.

Kidneys. Frequent passage of bloody urine, with extravasation in the loin.

In all cases, the portion of the body which received the brunt of violence must be considered, in determining what internal organs are probably injured.

How do you treat abdominal contusion with rupture of contained viscera?

Absolute rest. Opium.

If symptoms characteristic of internal hemorrhage, or rupture of a hollow viscus, appear, do an *exploratory laparotomy*. Bleeding from the liver or spleen can be checked by iodoform tamponade, or by the actual cautery. Torn vessels in the peritoneum can be ligated. Rents in the stomach or intestines can be united by sutures or brought to the surface. By irrigation, the peritoneal cavity can be freed of blood and extravasated matter. *Ruptured kidney* with lumbar extravasation should be treated by free lumbar incision and drainage.

What are the causes of traumatic peritonitis?

The bursting of an abscess, or the extravasation of urine,

blood, bile, or the contents of the alimentary canal into the peritoneal cavity.

Termination usually fatal, from collapse or blood poison.

What are the symptoms of traumatic peritonitis?

Severe pain, at first local, then general.

Extreme tenderness. Dorsal decubitus with legs and thighs drawn up. Breathing thoracic. *Abdomen distended and tympanitic;* later, dull in the flanks from effusion. *Obstinate vomiting. Complete constipation. Small, quick, wiry pulse.* Dry brown tongue. Temperature 103° to 104°.

In the septicæmic form there may be little pain or tenderness, and a normal or even subnormal temperature throughout

How do you treat traumatic peritonitis?

Prevent by absolute rest, cracked ice diet, hot fomentations, laparotomy.

Treat, on the development of the first symptom, by a full saline purge and turpentine enema. Open and wash out the peritoneal cavity. Insert a glass drainage tube. Stimulants and nourishment in teaspoonful doses.

Or treat expectantly, apply leeches to the abdomen, followed by hot fomentations or turpentine stupes. *Give opium* till the respirations are reduced to twelve in the minute.

How do you treat non-penetrating wounds of the abdomen?

Check all bleeding. Extensive extravasation may take place between the muscular planes if this precaution is not observed. *Pass sutures to the bottom of the wound*, approximating accurately. *Prevent tension* by position. Apply an antiseptic dressing, and a binder about the body.

If signs of inflammation appear, open freely (abdominal abscesses do not point). Guard against subsequent hernia.

Describe penetrating wounds of the abdomen.

These wounds involve the peritoneal cavity. There may be—
1. Simple penetration without visceral injury or protrusion.
2. Penetration with visceral injury, but no protrusion.
3. Penetration with visceral protrusion, but no injury.
4. Penetration with both protrusion and injury.

How do you treat simple penetrating abdominal wounds?

Thoroughly cleanse. Close the wound by sutures passed from within outward, including the peritoneum and the entire thickness of the abdominal wall. Apply an antiseptic dressing and a binder about the body, and place the patient in that position which will most effectually relax the wounded muscles. Give internally cracked ice for two days, then milk in small quantities. Opium, if indicated by pain or diarrhœa. If there has been hemorrhage into the peritoneal cavity, remove all blood by irrigation and insert a glass drainage-tube.

How do you treat penetrating wounds with visceral injury?

Enlarge, if necessary, and treat the visceral injury. Check bleeding from the liver and spleen by cautery, or iodoform tampons. Drain small wounds of the kidney. If the organ be extensively lacerated, do a nephrectomy. Wounds of the ureter require either a nephrectomy, or the formation of a urinary fistula by bringing the ureter to the surface. Wounds of the stomach or intestines should be sutured; if large, the sutured portion may be secured in the wound, the latter not being closed immediately (iodoform tamponade). Extravasation will then take place externally if the sutures yield. Slight punctures are closed by prolapse of the mucous membrane, and do not require suturing.

How do you determine as to the existence of a visceral injury in penetrating abdominal wounds?

If the wound is large, inspection and palpation may be sufficient.

In small wounds *intense pain* and *severe collapse*, with or without *escape of fæces, gas, bile, serum*, or *food*, indicate the nature of the injury.

Wounds of the stomach and intestines usually give a clear tympanitic percussion note over the liver.

In case of doubt inject hydrogen gas into the rectum; if the stomach or intestines are wounded, the gas will escape through the wound. Where there is no evidence of visceral wound treat as penetrating wound, performing an exploratory laparotomy on the first sign of internal hemorrhage or traumatic peritonitis.

How do you suture the intestine?

By the Lembert interrupted suture. The threads include *only the serous and muscular coats* of the bowel, are made of sterilized

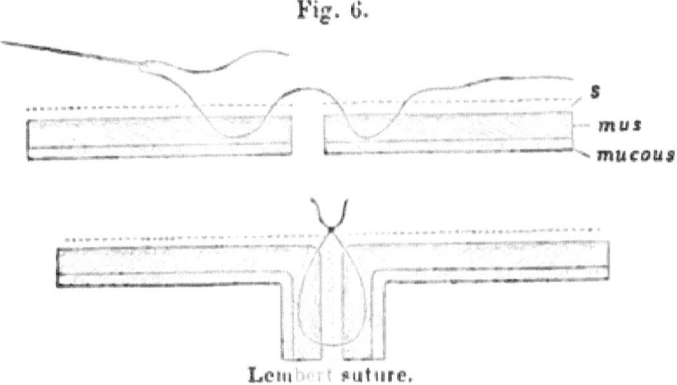

Fig. 6.

Lembert suture.

China silk, and are placed a twelfth of an inch apart. The suture is designed to approximate serous surfaces. It passes in and out on one side of the wound, across, and in and out on the other side, and is then tied. If the intestine is entirely torn across, or extensively injured, a portion may be resected, a V-shaped piece of mesentery removed, and the gut ends united by first bringing the peritoneal coat together by a circle of interrupted sutures, then invaginating the incision and approximating serous surfaces by Lembert's suture. This constitutes *Czerny's* suture. Or, the gut ends may be sutured through half their circumference, and the remaining opening secured in the wound, making an artificial anus.

How do you treat penetrating abdominal wounds with protrusion of viscera?

Carefully cleanse and return. If intestine is gangrenous, incise and leave in the wound; if congested and adherent, free from adhesions and return. The abdominal wound may be enlarged if necessary. Congested omentum should be ligated, removed, and the stump returned to the abdominal cavity. If the intestines protrude and are wounded, apply a Lembert suture and return, or make an artificial anus.

In all extensive injuries do not close the abdominal wound absolutely. Insert sutures, knot them loosely, and pack the wound with iodoform gauze. When danger from intra-peritoneal complications has passed away, approximate the granulating surfaces by removing the packing and drawing the sutures tight. The wound heals by secondary adhesion (third intention).

Describe laparotomy.

Preparation most thoroughly antiseptic. Incision in median line. Check all hemorrhage by hæmostatic forceps before opening peritoneum. The latter is nicked, held up by two fingers, and divided by scissors. Insert a large flat sponge to catch all oozing from wound. Irrigate the abdominal cavity, if necessary, with warm distilled water. If there is much shock, use hot water (not over 106). After the completion of the operation dry with sponges, inserting glass drainage-tube if there has been much manipulation or hemorrhage; close. First bring the peritoneum together with a line of interrupted catgut sutures; then insert some plate sutures of relaxation, using silk-worm gut. Suture together the fibrous investments of the two rectus muscles; finally unite the skin and subcutaneous tissues with interrupted sutures of approximation and continuous sutures of coaptation.

Dust with iodoform, apply a strip of protective, several layers of iodoform gauze, a thick investment of bichloride cotton, Mackintosh, and a moderately tight binder.

Give cracked ice for two days. Stimulants as required. *See that the bladder is regularly emptied*, drawing the water if necessary.

Describe tapping of the abdomen.

This operation is done for ascites.

See that the bladder is empty, pass a many-tailed bandage about the body to make pressure, let the patient sit up, leaning somewhat forward, make a skin incision in the linea alba, midway between the umbilicus and pubis, and thrust the trocar and canula into the abdomen. To avoid syncope draw off slowly,

Describe rupture of the bladder.

Cause. A blow or kick when the bladder is full. Fracture of the pelvis. Very rarely from simple over-distension. In retention from stricture the urethra more commonly gives way.

The rupture is usually vertical. Occurs more commonly in the posterior part, when the urine escapes into the peritoneal cavity, causing peritonitis. May occur in the anterior part, with extravasation into the loose cellular tissue of the pelvis, causing cellulitis with secondary peritonitis or septic poisoning.

What are the symptoms of ruptured bladder?

Collapse following an injury to the abdomen or pelvis, with absence of urine and presence of blood in the bladder, as demonstrated by passing a catheter. If the patient has passed his urine immediately before the injury, inject two ounces of warm boracic acid solution (4 per cent.) into the bladder; if there is an extensive rent in its walls, the solution will escape and cannot again be drawn off by a catheter. A catheter may sometimes be felt to pass through the rent.

How do you treat rupture of the bladder?

Do a supra-pubic cystotomy. If the rent is extra-peritoneal, insert a drainage tube. If the rent is intra-peritoneal, open the peritoneal cavity (through the same parietal incision), irrigate thoroughly to wash away all urine. Close the rent by the Czerny suture, taking particular care to see that no thread pierces the mucous membrane. Insert a drainage tube, tampon the external wound with iodoform gauze, and let the patient insure free drainage by the lateral decubitus.

These ruptures may be treated by the introduction and retention of a soft catheter passed through the urethra.

Burns and Scalds.

How are burns classified?

Burns are of six degrees.

1st Degree. Simple erythema followed by slight desquamation. There is no tissue destruction.

2d Degree. Vesication. The superficial layers of the epiderm are destroyed.

3d Degree. Destruction of the epiderm and the greater part of the true skin. A portion of the papillary layer, and the epithelium about the hair follicles and sebaceous glands escapes. This is of great importance in the subsequent healing, as skinning starts from these points as islands, and the elements of true skin are preserved to an extent. There is scarring, but not marked contractions. This is the most painful form of burn, from involvement of the nerve-endings.

4th Degree. Destruction of the skin and subcutaneous tissue. Scarring and contractions.

5th Degree. The deep fascia is penetrated and the muscles are involved.

6th Degree. Destruction of the entire part.

Describe the constitutional effects of severe burns.

Dependent on the *extent of surface involved*, and the depth.

Three stages.

1. *Shock and internal congestion.* Most marked in extensive burns of the trunk and head. The patient shivers and complains of cold.

2. *Reaction and inflammation.* Coming on in from one to two days. The patient complains of thirst and inflammatory fever. Internal congestion may run on to inflammation, causing meningitis, pleurisy, or peritonitis, according to the seat of the burn (head, chest, abdomen). Duodenal ulcer and nephritis are frequent complications.

3. *Suppuration and exhaustion,* setting in on the separation of sloughs. The patient often complains of cough and diarrhœa,

and may now perish from amyloid degeneration, exhaustion, or blood poison.

Great deformity ensues on cicatrization of deep burns.

What is your prognosis in severe burns?

Bad in burns involving one-third the surface, and in extensive burns upon the trunk. Fatal cases mostly perish within forty-eight hours from shock.

How do you treat burns?

Constitutionally. Treat the shock by external heat, hot bath, hypodermics of brandy, ammonia, atropia, and morphia. See that there is no retention of urine. When reaction and inflammation appear give a saline cathartic, neutral mixture. If the kidneys are congested apply dry cups, hot fomentations. Give liquid nourishment in small doses frequently repeated. Keep up the use of stimulants. Allay thirst by cracked ice. During the third stage give tonics and stimulants, push the nourishment, and treat diarrhœa by opium and astringents.

Locally. All burns beyond those of the first degree should be washed and dressed under all antiseptic precautions.

Burns of the second degree. Wash with 1 : 2000 sublimate solution, shave the surrounding skin, remove all loosened epithelium, wash again with 1 : 2000, using a soft brush or sponge for the injured surface, complete the cleansing with 1 : 5000 sublimate solution, cover with strips of protective wet in 1 : 5000, sprinkle iodoform *over* the protective, apply a thick layer of iodoform gauze overlapping the protective, a still larger and thicker layer of bichloride gauze, finally bichloride cotton and a bichloride bandage. Cure in ten days on removal of the dressing.

Burns of the third and fourth degrees, if limited in extent, are treated as burns of the second degree. Remove dressings when they become rank (ten days), thoroughly bathe in 1 : 5000, trim away sloughs, re-dress. When sloughs are all removed, and the burn converted to a granulating surface, skin graft.

When the burn is very extensive cleanse, wash, and remove loose cuticle as before, liberally sprinkle each region so treated with subnitrate of bismuth, cover with a single layer of lint or soft linen, held in place by one or two adhesive strips. Twice a day gently raise the edges of the lint, and sprinkle more bismuth wherever the coating has become loosened by discharge.

Or, puncture vesicles, but do not remove the cuticle, apply lint saturated in carron oil (lime-water and linseed oil in equal parts), and cover in with waxed paper and a light bandage. Change the dressing daily, *uncovering a small amount of surface at a time*, and redressing one part before another is exposed.

In extensive deep burns the continued warm bath may be employed till the sloughs separate.

Relieve the pain of burns of the first degree by white-lead paint.

Opium is indicated in all stages of severe burns.

FRACTURES.

What is a fracture?

The sudden solution in the continuity of a bone.

What are the causes of fracture?

1. Predisposing.
    - *a.* Local. Function, form, position, disease of the bone.
    - *b.* Constitutional. Includes conditions under which the bone becomes fragile, or subject to disease or injury—such as age, sex, rickets, locomotor ataxia, and necrosis.
2. Exciting.
    - *a.* External violence.
    - *b.* Muscular action.

What are the varieties of fracture?

Incomplete, partial, or greenstick. The bone is bent, but not entirely broken through. Stellate, grooved, and fissured fractures are also classed as incomplete.

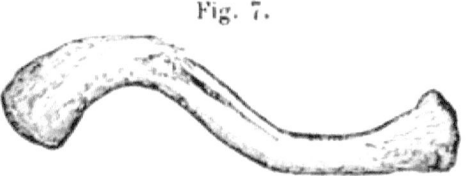

Fig. 7.

Greenstick fracture of clavicle.

Complete. The break involves the entire thickness of the bone.

Simple. Not accompanied by an open wound leading down to the break. A single uncomplicated fracture.

Compound. Accompanied by a wound leading down to the break.

Single. Having but one line of fracture, making in the long bones two fragments.

Multiple. Two or more fractures, the lines of breakage not communicating if these fractures are of the same bone.

Comminuted. The bone is broken into more than two pieces, the lines of fracture communicating.

Impacted. One fragment is driven into the other, and fixed in that position.

Complicated. Accompanied by an injury to some other important parts in the same region, as joints, bloodvessels, nerves, or muscles.

Further, fractures about joints are classed as:—

Intracapsular—within the capsular ligament.

Extracapsular—without the capsular ligament.

In young persons epiphyseal separation occurs, especially in the humerus, and constitutes epiphyseal fracture.

In what direction does the line of fracture extend?

It is generally *oblique*, but may be *transverse*, from direct vio-

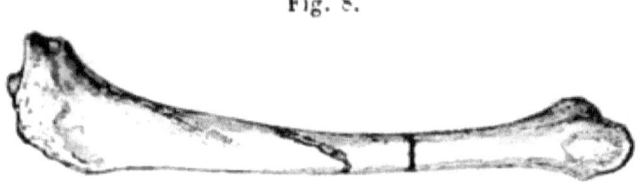

Fig. 8.

Oblique and transverse fracture of the tibia.

lence, *longitudinal*, when force is applied in the direction of the long axis of the bone, *spiral* or *stellate*.

What are the symptoms of fracture?

1. *Deformity* or displacement due to 1, the fracturing force; 2, the muscular contractility; 3, the weight of the part.
2. *Abnormal mobility.*
3. *Crepitus*, or harsh grating, both felt and heard on manipulation.
4. *Loss of function.*
5. *Pain and tenderness*, sharp and severe.
6. *Swelling and ecchymosis*, the latter appearing in certain lines.

What are the different kinds of displacement?

Angular or bending, *rotary* or *twisting*, *transverse*, *longitudinal* or overlapping.

When have you difficulty in recognizing displacement?

When but one of two parallel bones is broken, or when the short, flat bones are involved.

Under what circumstances is crepitus absent?

In greenstick and impacted fractures; when the fragments overlap considerably or are widely separated; when soft tissue is interposed between the ends of bone.

In epiphyseal fracture we have moist crepitus only.

What fractures do not present abnormal mobility?

Greenstick and impacted fractures.

How do you diagnose a fracture?

Deformity, unnatural mobility, and *crepitus,* if elicited, are absolutely diagnostic. If great swelling prevents a positive diagnosis, treat as a fracture till swelling subsides.

What is the general treatment of all fractures?

1. Reduce the fracture.
2. Retain it in position.
3. Treat inflammation and other complications, either constitutional or local.

How do you reduce a fracture?

1. By *extension* or traction, made by the surgeon steadily pulling upon the lower fragment.
2. *Counter-extension* or fixation of the upper fragment.
3. *Coaptation* or adjusting the broken ends of the bone to their proper position.

How do you overcome muscular spasm?

If muscular spasm interferes with reduction, it must be overcome by *position, etherization,* or *tenotomy.*

How do you retain the bones in proper position?

By means of splints and bandages. Splints may be made of wood, tin, gutta-percha, binders' board, leather, etc.

Bandages may be made of muslin, linen, or gauze, or may have incorporated with them various materials which, hardening, make a solid and firm dressing, as plaster, silicate of potassium, gum, etc.

Under what circumstances are the fixed dressings, as plaster, applied?

Primarily, in fractures attended with little swelling, displacement, or damage to the soft parts. Secondarily, in fractures of the lower extremity, after the subsidence of swelling and inflammatory symptoms.

What rules guide you in the application of splints?

1. Splints should be well padded.
2. They should fix the joints above and below the break.
3. The extremities of the limbs should be left exposed to view (fingers and toes).

Circular compression must be avoided, primary rollers being absolutely discarded in fractures of the leg or forearm. Applied with great caution in fractures of the thigh or upper arm.

How often do you re-dress a fracture?

The fracture dressing must be inspected daily for one week. If too loose or too tight, or if there is evidence of displacement, the dressing must be renewed. Otherwise, twice weekly will be sufficient.

What complications may arise, and how should they be treated?

1. *Œdema and swelling often accompanied by blebs.* Treat by loose bandaging at first, and evaporating lotions; follow by pressure.
2. *Ulceration and sloughing of soft tissues.* Free ulcerating spot from pressure by careful padding of splint.
3. *Muscular spasm.* Treat by moderate pressure, morphia injections, or tenotomy.
4. *Gangrene.* Usually the result of too tight dressing, or laceration of main artery. Relieve pressure.

Rarely. Venous thrombosis, embolism, fat embolus—causing death by asphyxia. Treatment: cardiac stimulants.

How do you treat compound fractures?

If the external wound is small and the fracture not otherwise complicated, thoroughly cleanse with bichloride 1 : 1000, and close with absorbent cotton saturated in a solution of ether, iodoform, and collodion, equal parts of each. Splint as usual. If inflammatory symptoms arise, or if there be much original comminution or laceration of soft parts, pick out loose fragments, thoroughly cleanse, irrigate with bichloride solution 1 : 1000, drain, and apply antiseptic dressing, splinting as usual. If wound be older than twenty-four hours, wash with 1 : 5 carbolic solution (acid carbol. 1, alcohol 5).

What complications arise in the treatment of compound fractures?

Necrosis, osteomyelitis, periostitis, extensive sloughing of soft tissues.

What is the pathology of fracture?

There is first free *bleeding* from the vessels of the injured bone, medulla, and surrounding soft parts. This is followed by *inflammation* with exudation, absorption of blood clot, and deposit of plastic lymph about the seat of injury. *Organization* completes the process; the plastic lymph is converted first into cartilage, then into bone.

What is callus?

The *plastic* lymph which is organized into bone tissue for the repair of fractures.

How is the callus disposed about a fracture?

A portion is deposited as a fusiform swelling ensheathing the two broken bone ends, called *ensheathing callus;* a portion fills the medullary canal above and below the break acting as a supporting pin, called pin or *central callus*. A portion is directly between the broken surfaces restoring their continuity, called *intermediate* or *definitive* callus.

What is meant by temporary and permanent callus?

The ensheathing and pin callus is temporary, being absorbed when the bone is firmly united by the intermediate or permanent callus.

What period of time is occupied by the various processes necessary for the repair of fracture?

Absorption of clot first week, formation of plastic lymph and beginning organization second week, ossification of the callus 4 to 8 weeks, absorption of temporary callus one year.

What complications are common to all fractures?

Shock.

Retention of urine, treat by catheter.

Traumatic delirium, especially in drunkards—sedatives, stimulants.

Hypostatic congestion of lungs.

What compound fractures require amputation?

Compound fractures associated with—

1. Very extensive laceration of soft parts.
2. Great destruction of bone substance.
3. Injury to the main artery of leg or thigh (femoral or post-tibial).
4. Injury to knee or ankle, if extensive.

Define delayed union and non-union.

Union is delayed when fractures are not firmly joined by callus in 4 to 6 weeks.

We have *non-union* or *ununited fracture* when the continuity of the bone is not restored after twelve weeks.

What are the causes of delayed union and non-union?

1. *Constitutional* include all conditions depressing to health and nutrition, as acute fevers, syphilis, phthisis, scurvy, nephritis, etc.
2. *Local.* *a. Undue mobility of fragments* often from improper splinting or meddlesome interference;

   b. Separation of fragments, by muscular action, or by interposition of soft parts or necrosed bone.

   c. Interference with blood supply, as in intracapsular fracture.

How do you treat non-union?

Treat constitutional conditions.

Locally the means adopted would be in the order given below, one failing the next should be tried. The object of all these methods is to set up an acute aseptic inflammation, which shall provide sufficient exudation for the formation of healthy callus.

1. *Absolute fixation*, careful dressing, plaster bandage.
2. *Friction*. Rub ends of bone together either manually or by getting patient up and allowing some use, the fragments being held in apposition by fixed plaster bandage or apparatus.
3. *Drill* fragments subcutaneously to excite inflammation and deposition of plastic lymph; treat subsequently by absolute fixation.
4. *Drill and pin fragments together* leaving the pin in place.
5. *Resection* of the ends of the bones, joining the fresh surfaces by silver wire. Drain thoroughly and close the wound. Secure fixation by careful splinting.

Name the forms of non-union?

1. No union whatever between the fragments.
2. Ligamentous union.
3. False joint.

What is vicious union?

Union accompanied either by great deformity, or by the binding together of bones which should move on each other, as the radius and ulna.

How do you treat vicious union?

If recent, restore immediately by force, or by splints and pressure. If firm union has taken place, or the fracture is not amenable to other treatment, the bone should be broken again, properly set, and fixed in position. Deformity from exuberant callus gradually disappears. Should it persist, and should pressure symptoms arise, callus must be cut away.

How do you treat an injury which you suspect may be a fracture?

Treat as a fracture, subsidence of swelling will clear the diagnosis.

Under what circumstances do you use anæsthetics in the diagnosis and treatment of fracture?

1. In case of difficulty or doubt.
2. In complications requiring prolonged or painful manipulations.
3. Where reduction is not readily effected.

How do you treat the swelling and ecchymosis common to all fractures?

Evaporating lotions for two or three days, followed by carefully guarded pressure. Four ounces of alcohol and four drachms of ammonium muriate, two ounces of the solution of acetate of lead, or eight ounces of laudanum, to the pint of water. Apply on lint which must not be covered with oiled silk, but kept constantly wet by the solution.

What is the cause of the late discoloration in fractures?

The effused blood gradually works its way to the surface, between layers of fascia, in the path of least resistance; the disintegration of the red corpuscles causes the ecchymosis or discoloration.

Special Fractures.

Describe fractures of the nasal bone.

Cause. Direct violence.

Signs. Displacement, backward or lateral. Crepitus. Unnatural mobility. Deformity. Very rapid swelling. Free bleeding.

How may this fracture be complicated?

1. Profuse hemorrhage.
2. Emphysema of surrounding soft parts.
3. Deflection or fracture of septum nasi.
4. Injury to base of brain through the perpendicular plate of ethmoid.

Give the treatment of fracture of the nasal bone.

Reduce at once by pressure exerted by a director or closed hæmostatic forceps passed into the nostril. Retain in place, if necessary, by packing the nostrils with iodoform gauze or an inflatable rubber bag, the respiratory tract being kept open by a rubber tube. If there is much comminution and these means fail, *fasten the fragments together with pins*, passed from the outside, taking in the periosteum. Inspect the nostrils for deflection of septum, which must always be replaced.

Check hemorrhage by heat, cold, astringents, or packing.

Treat swelling by evaporating lotions.

Always reduce thoroughly.

Describe fractures of the superior maxillary bones.

Ordinary fracture symptoms, generally accompanied by great swelling.

Common seat of fracture, alveolar process—at times nasal process, malar process, or body of maxilla. The anterior wall of the antrum may be driven in.

How do you treat fractures of the superior maxilla?

Reduce, if deformity. If the bone is driven in, raise by pressure applied from the mouth, or by means of an elevator passed through a small skin wound. Retain alveolar process by making the lower jaw the splint, applying a Barton's bandage; treat swelling and inflammation by evaporating lotion, applied on lint (alcohol and water equal parts).

Describe fractures of the inferior maxilla.

Usual seat. Near or through the anterior mental foramen.

Fractures also occur at the *symphysis;* through any part of the *body;* through the *ramus;* through the *condyloid* process; through the *coronoid* process.

These fractures are often compound, from rupture of the mucous membrane.

Give the symptoms of fracture of the inferior maxilla.

Body. The cardinal signs of fracture, together with pain, swelling, dribbling of saliva, disability. The central portion of the bone is pulled downward and backward by the digastric, geniohyoid, and geniohyoglossus muscles.

Fractures of the ramus give little deformity, the bone being held in place by the masseter without, the internal pterygoid within. Manipulation elicits mobility and crepitus.

In fractures of the neck, the condyle is pulled forward and inward by the external pterygoid, causing great pain and crepitus on opening or closing the mouth.

Give the treatment for fracture of the inferior maxilla.

Careful reduction and the application of a moulded pasteboard splint, well padded with cotton, and held in place by a Barton's or Gibson's bandage. Frequently wash the mouth with saturated solution of boracic acid.

If the dressing fails to keep the fragments in proper position, they should be drilled and wired in place. The dressing can be removed in five weeks.

Give the symptoms of fracture of the hyoid bone.

Seat of injury. Greater horn. Pain on eating or speaking, together with the cardinal signs of fracture, elicited by examining with the fingers of one hand in the pharynx, while the other hand outlines the bone from without. The displacing factor is the middle constrictor.

Give the treatment for fractures of the hyoid bone.

Reduce by pressure, keep the head between flexion and extension, support by a pasteboard collar, give nutrient enemata for four days, then, if dysphagia be still great, feed by the œsophageal tube.

Give the symptoms of fracture of the laryngeal cartilages.

Usual seat. Thyroid cartilage. *Symptoms*—Aphonia, dyspnœa, and *bloody expectoration*, together with *emphysema, deformity,* and *possibly moist crepitus.*

Treatment. On the appearance of dyspnœa, intubation, or, that failing, tracheotomy. Feed by rectum for some days, and secure absolute rest to the parts.

Describe fractures of the clavicle.

Cause. Usually indirect violence, as falls on the palm of the hand.

Seat. May be any portion of the bone, *generally* outer portion of middle third.

Direction. Oblique.

Displacement. Shoulder falls downward, forward, and inward, shortening detected by measurement from middle of upper border of sternum to coracoid process.

What causes the displacement in fractured clavicle?

The outer fragment drops downward, inward, and forward from the weight of the shoulder, and the action of the two pectorals, the latissimus dorsi and the serratus magnus; the inner extremity of the outer fragment is thrown somewhat backward by the rhomboidei and levator anguli scapuli, so that it lies behind and below the outer extremity of the inner fragment, which is slightly tilted up by the sterno-cleido mastoid.

Give the symptoms of fractured clavicle.

Crepitus and preternatural mobility readily elicited by pushing up and rotating the humerus.

Deformity detected by passing the finger along the subcutaneous surface of the bone, by inspection, by measurement; shoulder flattened, arm disabled.

Fractures of acromial and sternal end necessarily allow of but little displacement. If external to conoid and trapezoid ligaments, there is marked displacement of the outer fragment.

Give the treatment for fractured clavicle.

The object of the treatment is to restore the fragments to their proper position by forcing the shoulder upward, outward, and backward. This is accomplished by—

1. *Sayer's dressing.* Strips of adhesive plaster three and one-half inches wide. The first is long enough to surround the body including the arm. This strip encircles the arm over the insertion of the deltoid in the form of a loosely fitting loop, which must be made secure by sewing. Draw the arm somewhat downward and backward, to make tense the clavicular origin of the pectoralis major, and fasten it in this position by carrying the strip entirely around the body securing it to itself in the back.

The second strip begins at the sound shoulder, is carried obliquely over the back to the elbow of the injured side, which is received in a slit provided for the purpose, it is then carried upward across the front of the chest to its point of origin. This forces the shoulder upward, backward, and, by pulling the elbow in, also outward.

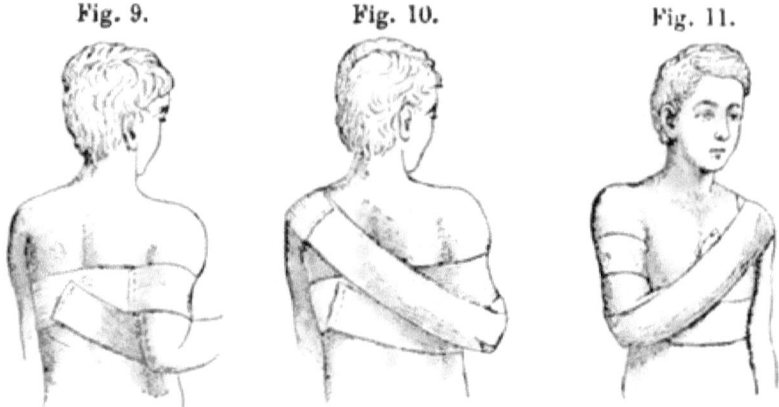

Fig. 9.   Fig. 10.   Fig. 11.

2. The *recumbent posture*, supine, with the arm carried across the chest, is the best theoretical treatment for this injury.

3. *Velpeau's dressing.* A pad fastened in the axilla of the injured side. The forearm flexed on the arm and carried across the chest till the hand rests on or near the sound shoulder. Careful manipulation of the fragments into proper position, and the application of Velpeau's bandage.

4. *Désault's dressing.* A pad fixed in the axilla by the first roller. The arm bound to the side by the second roller. The shoulder pressed upward and backward by the third roller.

Union in about four weeks; carry the arm in a sling for one or two weeks longer.

Describe fractures of the scapula.

Cause of fracture. Direct violence.

Seats of fracture through 1. Body or inferior angle. 2. Surgical neck (supra-scapular notch). 3. Glenoid cavity. 4. Acromion or coracoid processes.

What are the symptoms of fractured scapula?

In all situations there are found disability, pain, swelling, crepitus, and preternatural mobility.

Neck (through suprascapular notch). Disability complete. If conoid and trapezoid ligaments are torn there will be a space between the acromion and humerus- disappearing on pressing the arm upward, but recurring again when the support is removed. Coracoid process moves with humerus, the acromion remains fixed.

Acromion process. If behind the acromio-clavicular articulation the shoulder is flattened, and drops downward, forward, and inward. Crepitus and undue mobility.

Coracoid process. Complete disability. Unnatural motion may be felt by pressing a finger deeply in the region of this process and pushing up the elbow.

Give the treatment for fractures of the scapula.

Body. Compress to both borders of the scapula, adhesive plaster extending circularly from the spine to the sternum, Velpeau or Désault bandage, with the arm vertically to the side.

Neck, glenoid cavity, acromion or coracoid process. Towel in axilla, and Velpeau or Désault bandage.

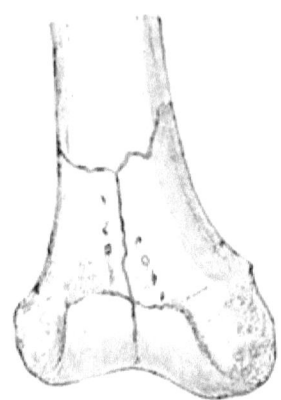

Fig. 12.

Comminuted or **T** fracture.

Describe fractures of the humerus.

Muscular attachments.

To *greater tuberosity.* Supraspinatus, infraspinatus, and teres minor.

To *lesser tuberosity.* Subscapularis. *Anterior bicipital ridge.* Pectoralis major. *Posterior bicipital ridge.* Latissimus dorsi, teres major. *Shaft.* Coraco-brachialis, deltoid, triceps. *Internal condyle.* Pronator radii teres and common flexor tendon. *External condyle* and *condyloid ridge.* The two supinators, anconeus, extensor carpi radialis longior, and the common extensor tendon.

There may be fractures of the head, anatomical neck, tuberosities, surgical neck, including epiphysis, shaft; there may be supra-condyloid, inter-condyloid, **T** or comminuted, condyloid, epicondyloid (internal only) fractures.

Give the symptoms of fractured humerus.

In all, except the impacted fractures of the anatomical neck, there are pain, crepitus, preternatural mobility, deformity, disability, and swelling.

Head and anatomical neck. Symptoms obscure, slight shortening, crepitus on upward pressure and rotation, broken extremity may be felt in axilla.

Greater tuberosity. Depression under acromion process, widening of shoulder, smooth bony prominence (head of bone) under coracoid, crepitus on rotation and pressing tubercles together, external rotation cannot be performed by the patient.

Surgical neck. (That portion of the shaft of the humerus lying between the tuberosities and the insertion of the latissimus dorsi and teres major muscles.) Commonest seat of fracture. Direction transverse. Shortening (measured between acromion process and external condyle). Lower fragment drawn inward and forward by latissimus dorsi, pectoralis major, and teres major, pulled upward by deltoid, biceps, triceps, and coraco-brachialis. Rough end of lower fragment felt near coracoid process. Unnatural mobility and crepitus on extension and rotation.

Epiphyseal. As in surgical neck, except that it occurs in young people, and that the crepitus is moist and the fragments smooth.

Shaft of humerus. Mostly below middle third. *Direction* oblique. *Deformity*, overlapping, from biceps and triceps; if above insertion of the deltoid the lower fragment is pulled outward by that muscle; if below, the upper fragment is tilted forward. Cardinal signs of fracture readily detected.

Supra-condyloid. Projection in front and behind. That in front is due to the rough end of the upper fragment; that behind is due to the condyles and olecranon *occupying their normal relation in regard to each other*. Shortening between acromion process and external condyle. Reduction easy, but deformity promptly recurs.

Intercondyloid. Increased breadth between the condyles, and crepitus elicited by pressing and rubbing them together.

Condyloid. Crepitus and mobility on manipulating the bony prominences, displacement slight.

All fractures about the elbow-joint are accompanied by great and rapid swelling.

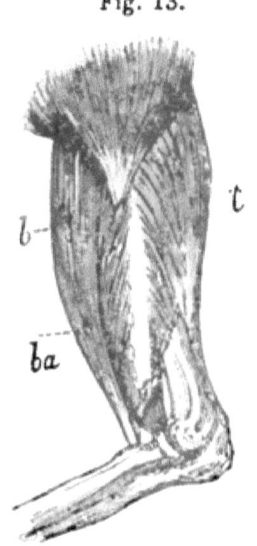

Fracture of the lower extremity of the humerus.

Dressing for fracture of the upper third of the humerus.

Give the treatment for fractures of the humerus.

Upper extremity. Including intra- and extra-capsular, trochanteric, and fractures of the surgical neck.

Fasten a folded towel in the axilla by a bandage and adhesive strap.

Flex the arm, and carry the elbow slightly forward, apply a spiral reversed from the hand to the seat of fracture. Place a moulded pasteboard cap, or three straight, narrow, external splints, reaching from the acromion process to the external condyle, upon the outer aspect of the arm and shoulder, bind in place by a few circular turns of a roller, and complete the dressing by fastening the arm to the side, and slinging the forearm at the wrist.

Shaft of humerus. Primary roller up to the seat of fracture, well padded internal angular splint, avoiding pressure upon internal condyle, shoulder cap extending to external condyle or below on forearm, arm bound to the side by circular turns of the roller, and slung at the wrist.

If obstinate deformity from outward tilting by the deltoid, relax by dressing in the abducted position for a few days.

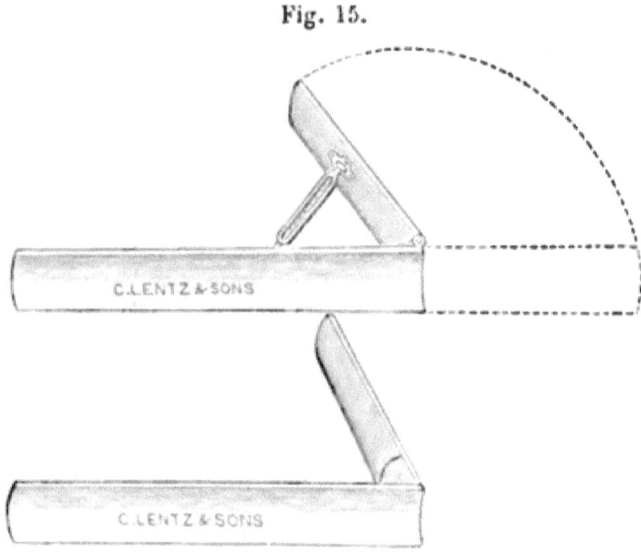

Fig. 15.

Anterior angular splints.

Supra-condyloid. Internal angular and external moulded splint, or anterior angular splint and posterior moulded trough.

Condyloid. Very obtuse angled, anterior, or internal splint.

What complications may arise in the treatment of these fractures?

1. Non-union, always in intracapsular fractures, frequently in fractures of the shaft.

2. Paralysis, from injury to the musculo-spiral or ulnar nerves.

3. Anchylosis, from inflammation within or about the joints, particularly the elbow.

How do you avoid anchylosis in fractures about the joints?

By practising passive motion. Begin in four weeks for the shoulder-joint; one week for the elbow. Promptly treat inflammation by cold, local depletion, aspiration at times, and pressure.

How long do you continue treatment?

Five to eight weeks, replacing the splints with a sling in that time.

What fractures occur in the ulna?

Seats of fractures: shaft, olecranon, styloid or coronoid processes.

Cause, direct or indirect violence. *Usual seat* lower third.

Give the symptoms of fractured ulna.

Cardinal symptoms as in all fractures.

Shaft, being subcutaneous, deformity, crepitus and undue mobility readily recognized.

Olecranon. Loss of power to extend, undue mobility; crepitus on extending forearm and pressing olecranon in position. Displacement often very slight. If aponeurosis is torn through, the process is drawn well up the arm from between the condyles, leaving a perceptible gap.

Coronoid process. Very rare. Tendency to backward luxation of ulna, movable bony prominence in front.

Styloid process. Mobility. Crepitus detected by carrying hand towards radial border.

Give the treatment for fractures of the ulna.

Olecranon. Figure-of-eight about the joint, the upper segment looping behind the displaced fragment, pulling it downward. Application of a very obtuse anterior or internal angular splint.

Shaft. Two well padded splints, each wider than the forearm, one reaching from the internal condyle to the tips of the fingers, the other from the external condyle to the metacarpo-phalangeal articulation. Reduce the fracture, apply splints, with the hand midway between pronation and supination. Support the forearm through its whole extent by a handkerchief.

Coronoid process. Anterior angular splint and compress. Passive motion in three weeks.

Fig. 16.

Dressing for fractures of one or both bones of the forearm.

Styloid process. Reduce, apply a compress. Bandage to a Bond splint, or apply anterior and posterior straight splints.

Describe fractures of the radius.

Seats of fracture. Head, neck, shaft, lower extremity. Ordinary seat, *lower extremity.*

Muscular attachments. Biceps, supinator brevis, pronator radii teres, pronator quadratus, supinator longus.

What fractures occur at the lower extremity of the radius?

Barton's (rare). A chipping off of the posterior lip of the articular surface.

Colles's. Common. A transverse break ¼ inch to 1½ inches above the joint.

Smith's. A transverse fracture 1½ inches to 2½ inches above the joint.

FRACTURES. 123

Fig. 17.

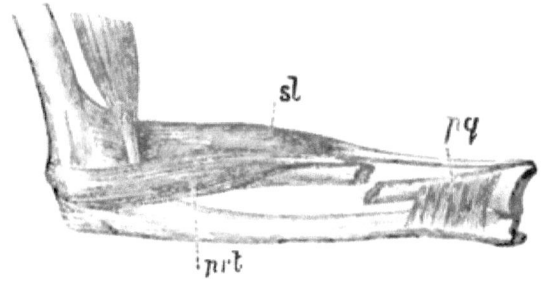

Give the symptoms of fractured radius.

Cause. Fall on the palm of the hand. Direct violence.

Lower extremity. Silver fork deformity. Lower fragment lies posterior to the upper fragment. Hand carried towards radial side by supinator longus, extensor carpi radialis, and extensors of the thumb. Crepitus and mobility on rotation. All symptoms marked.

Shaft. Upper fragment slightly tilted forward by biceps, and, if above insertion of pronator radii teres (middle third), supinated by biceps and supinator brevis. Lower fragment pronated by two pronator muscles, tilted towards ulna by pronator quadratus and supinator longus. If below the insertion of the pronator radii teres, deformity as before, except that both fragments are midway between pronation and supination. Crepitus and mobility elicited by rotation.

Neck of radius. Upper fragment supinated by short supinator, lower fragment pulled forward by biceps. Crepitus, mobility, and deformity detected by pressing the thumb into the bend of the elbow and rotating the forearm.

Both bones. Usual seat lower third. Shortening and angularity often marked. Crepitus, unnatural mobility by grasping the bones on either side of the fracture and manipulating, or by placing the thumb upon the head of the radius, making extension, and rotating.

Upper fragments pulled forward by biceps, brachialis anticus, and pronator radii teres. Lower fragments approximated by pronator quadratus; overlapping from the action of the flexors and extensors.

How do you treat fractures of the radius?

Neck. Anterior angular splint, and compress over upper end of displaced shaft. Dress in *supination*.

Shaft. As for shaft of ulna. Reduce by extension, counter-extension, manipulation.

Lower extremity. Reduction most important. Fragments once placed in proper position usually remain so.

Fig. 18.

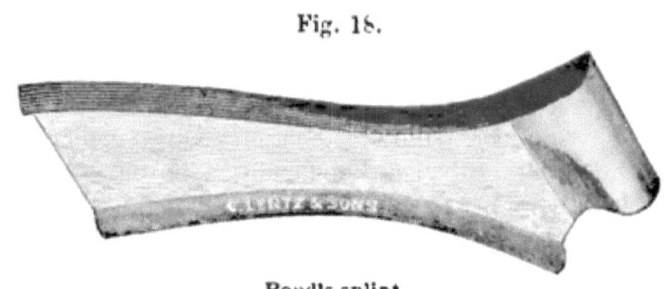

Bond's splint.

Reduce thoroughly by extension, pressure, and manipulation. Apply a Levis or a Bond splint, or simply circular strips of adhesive plaster. In all cases leave the fingers free, and encouraging their use. The Bond splint requires two pyramidal pads, the base of the posterior one to go over the upper extremity of the lower fragment, the apex pointing toward the fingers. The base of the anterior one to go under the lower extremity of the upper fragment, the apex pointing toward the elbow. Firm union in five to seven weeks.

Fractures of both bones, or shaft of either, including Colles's fracture, complicated by a fracture of the styloid process of the ulna.

Two straight splints wider than the forearm, as in fractures of the shaft of the ulna.

Sling all fractures of the forearm by means of a handkerchief supporting it throughout its entire extent.

What forearm fractures are dressed in supination?

Dress fractures above the insertion of the pronator radii teres with the palm up; in all other fractures, dress with the thumb up (midway between pronation and supination).

Describe fractures of the metacarpus.

Usually second or fifth. Posterior angular projection, from distal end of bone being pulled forward by the flexors. Crepitus and mobility elicited by seizing and manipulating the two extremities of the bone.

Give the treatment for fractures of the metacarpus.

Treat by an anterior splint to the hand and forearm, padding well to preserve the concavity of the palm. Compress posteriorly if any tendency to deformity. Retain the dressing for five weeks. Passive motion in three days.

Describe fractures of the phalanges.

Rare. Due to direct force; readily diagnosed by manipulating the finger bones. Treat by anterior moulded, posterior straight splint, extending to the wrist. A long palmar splint may be used.

Describe fractures of the pelvis.

Cause. Great and direct violence.

Seats. Crest of ilium, basin of pelvis, acetabulum, sacrum, or coccyx.

Symptoms. In all these fractures there is a sense of falling apart.

Crest. Patient leans toward the affected side; crepitus and mobility on grasping and manipulating the bone. External evidence of injury, discoloration, swelling, etc.

Pelvic basin. Crepitus and mobility may be elicited by grasping the iliac spines and attempting to move them in opposite directions; great pain, and inability to sit or stand; often a line of ecchymosis along Poupart's ligament and the crest of the ilium. Examination per rectum or vagina may reveal displacement or crepitus.

Acetabulum. Either the floor or the rim may be fractured; caused by blows on the trochanter.

Floor. Great pain on attempting to stand, or in any way moving the femur; crepitus best detected by thrusting the femur directly upward; very slight shortening.

Rim. Usually the upper and posterior part is broken off.

Subluxation of femur backward. On circumduction, the head of the bone can be felt to slip out at a certain point, returning to its proper position as the motion is continued; there is crepitus.

Sacrum and coccyx. Direction transverse. Cause, direct violence. There may be some anterior projection from the action of the coccygeus and levator ani muscles. Crepitus and mobility, detected by a finger in the rectum. Pain on defecation.

How are these fractures treated?

Place the patient on a fracture bed, *i. e.*, a firm, hard, evenly padded bed, with a central perforation through which the contents of the bowel may be passed without moving the patient. Apply a broad bandage or binder tightly about the pelvis; tie the knees together. The most comfortable position is usually on the back, with the thighs and knees flexed, and supported by pillows; allow the patient to assume the position of his choice. If there is displacement of the coccyx, pack the rectum with iodoform gauze or an inflated rubber bag.

Fractures of the acetabulum are treated by extension, and sand bags or splints, as fractures of the femur.

Describe fractures of the femur.

Muscular attachment—

To greater trochanter—Two gluteals (medius and minimus), two obturators, two gemelli, pyriformis, quadratus femoris. All external rotators except the glutei.

Lesser trochanter—Psoas, iliacus (below), both flexors and external rotators.

Condyles—Gastrocnemius, plantaris, and popliteus.

Seats of fracture. Neck—Intracapsular, extracapsular, mixed. *Shaft. Lower extremity*—Supracondyloid, intercondyloid, **T** or comminuted, and condyloid.

Give the symptoms of intracapsular fracture of femur.

Occurs in aged people, frequently females, from slight violence. Hip flattened, trochanter less prominent, and lying nearer to the anterior superior spinous process of the ilium, with its upper border above Nélaton's line (a line from the anterior superior iliac spine to the tuberosity of the ischium).

FRACTURES.

Crepitus elicited by pressure upon the trochanter, and making traction and internal rotation. Pain on motion. Preternatural mobility, foot can be everted till the heel looks directly upward. Swelling not accompanied by marked ecchymosis. Shortening from ½ to 1½ inches; may be slight at first and progressively increase. Loss of power.

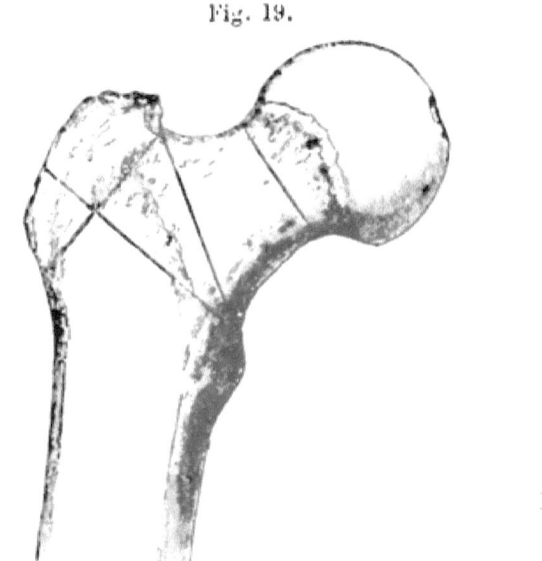

Fig. 19. Lines of fracture of the upper extremity of the femur.

Fig. 20. Intracapsular fracture of the neck of the femur.

Give the symptoms of extracapsular fracture of the femur.

Cause. Considerable direct violence. It occurs in middle-aged males, with well-marked external evidence of injury, *i. e., swelling* and *discoloration.*

Crepitus distinct, harsh, readily elicited.

Shortening marked, 1 to 2½ inches.

Give the symptoms of impacted fracture of the hip joint.

The impacted fracture may be either intra- or extracapsular. There will be: 1. No crepitus. 2. Slight shortening, not disappearing on traction. 3. Loss of function in the limb, but not absolute. 4. Evidence of much injury to the soft parts.

The foot may be inverted or everted.

Give the symptoms of fracture of the great trochanter.

This injury often accompanies extracapsular fracture, but may exist alone. *Cause.* Direct violence. It is characterized by pain, swelling, discoloration, and crepitus. Unnatural mobility elicited by pressing into place the broken fragment, which may be felt as a hard lump upon the dorsum of the ilium.

Give the symptoms of fracture of the shaft of the femur.

Cause. Direct violence.

Common seat. Middle third. *Direction* Oblique.

Eversion of foot, very marked; shortening, increased mobility, crepitus, loss of power. Upper fragment, especially in the upper third, drawn forward and everted by psoas, iliacus, and external rotators; lower fragment pulled up and in by adductors, flexors, and extensors.

Give the symptoms of fracture of the lower extremity of the femur.

Supracondyloid. Lower fragment pulled back by gastrocnemius, shortening, and eversion.

Intercondyloid, condyloid, or **T** *(transverse and intercondyloid).* Increased measurement between the condyles, associated with great and rapid swelling of the knee. Undue mobility and

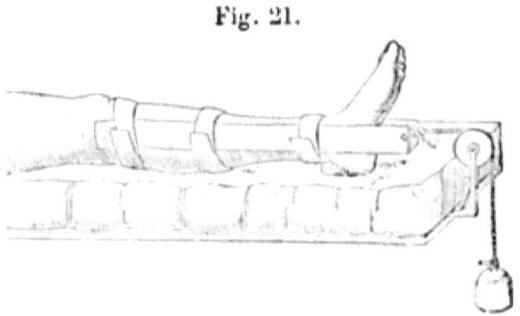

Fig. 21.

Extension applied for fracture of the femur.

crepitus, elicited by bending the knee, or by grasping the condyles and pushing them in opposite directions. Very great pain.

FRACTURES.

How do you treat fractures of the femur?

Upper extremity and shaft. Extension by adhesive plaster 2½ inches wide and long enough to extend from the upper end of the lower fragment, on both sides of the limb, and leave a 4 to 6 inch loop hanging free below the sole of the foot; in this loop is laid a piece of thin splint board 2½ inches wide, and so long, that when traction is made, the plaster will stand free from the malleoli. This board is fastened in place, and through a hole in its centre a cord or bandage is passed. The adhesive plaster is placed along the inner and outer aspect of the limb up to the seat of fracture, and secured in place by a few strips carried around the limb, and a neatly applied spiral reversed bandage of the lower extremity. After an hour or two the plaster is

Fig. 22.

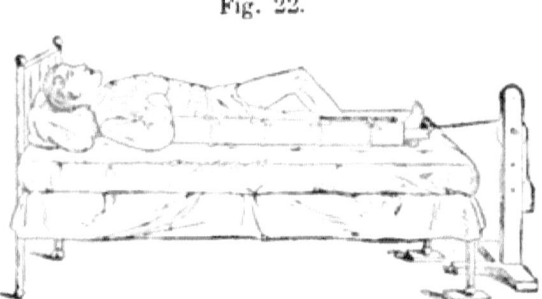

Dressing for fractured femur.

tightly adherent, when the extending cord is carried over a pulley, a weight is attached, and a pad of oakum is put beneath the tendo Achillis. A sand-bag, or a bran-bag and straight splint is placed on each side of the leg, the inner extending from the sole to the perineum, the outer from the sole to the axilla, and the foot of the bed is raised two to four inches to provide for counter-extension. The position of the foot is *slight* eversion, and flexion. The inner borders of the inner malleolus, internal condyle, and ball of the great toe should lie nearly in the same vertical plane, the great toe pointing directly upward.

Fractures of the upper extremity or shaft of the femur may also be treated by well-padded straight internal and external splints. The shaft may be treated by plaster or other fixed bandage, or by straight short splints buckled about the seat of fracture. In all cases, except in impacted fracture, extension should be used.

What dressing should be applied when the upper fragment projects anteriorly?

Relax the psoas and iliacus by flexing the thigh and supporting it and the leg upon a double inclined plane, raise to such an angle that the deformity is corrected. Apply the extension plaster from the knee to the upper end of the lower fragment, make a stirrup as before, then carry the extending cord over a pulley, so elevated that traction is made in the long axis of the femur.

Give the treatment for fractures of the great trochanter.

A bandage about the hips with a moulded cap to keep the trochanter in position, and a long straight external splint extending from the axilla to sole.

How do you treat fracture of the lower extremity of the femur?

If there is obstinate angular deformity, section of tendo Achillis. If marked shortening, extension as before, carried not quite up to the seat of fracture. A splint, or long fracture-box, well padded with pillows, should be used. Evaporating lotions, or aspiration, for accompanying synovitis.

How long should treatment be continued in fractures of the femur?

Treatment, five to eight weeks. Passive motion of the knee joint after fourteen days. Massage before allowing the patient to put the leg down. Application of plaster, or other fixed dressing about the fracture, before walking is allowed.

How do you treat fracture of the femur in infants?

Reduce by extension, counter-extension, manipulation. Place in position a carefully padded external splint extending from the axilla to the sole of the foot, and fasten it in place by a silica or plaster dressing. Treatment for four weeks.

How do you distinguish between intracapsular and extracapsular fractures of the femur?

In extracapsular—

1. Crepitus is rougher, more readily elicited, and feels as though immediately beneath the fingers of the surgeon.

2. Swelling and discoloration are greater and more immediate.

3. Deformity or shortening is more marked, but eversion cannot be carried so far as in intracapsular fracture.

4. On rotation the trochanter is found to pass through an arc of less radius in extracapsular fractures.

Describe fractures of the patella.

Causes. Direct violence, and muscular action.

Direction. Transverse or longitudinal. Generally, but not always, marked separation of fragments.

Give the symptoms of fractured patella.

Power of extension lost. Gap between fragments, increased on flexion. Great swelling. In longitudinal fractures, crepitus and mobility on grasping the two sides of the bone and pressing in opposite directions.

How do you treat fractures of the patella?

If there is not much separation, elevate and apply a straight posterior splint to the thigh and leg. If great swelling, cold and evaporating lotions for one or two days, aspirating the joint if necessary. The posterior straight splint is provided with lateral pegs and ratchets, to which are attached strips of adhesive plaster which are looped over the upper and lower fragments; by turning these pegs, the lower fragment is steadied, and the upper fragment is drawn down in position. Fix the lower fragment first, then the upper. Imbricate the plaster strips from above downward. If the edges of the fragments tilt forward, carry a piece of strapping transversely around the limb. Complete the dressing with a figure-of-eight bandage. Begin passive motion in two or three weeks. Continue the splint for six or eight weeks. Follow with a stiff bandage, plaster or glass, and keep the patient on crutches for several months.

These fractures may also be treated by Malgaigne's hooks applied under strict antiseptic precautions. Or by making a transverse incision, clearing the breach between the fragments and the knee joint of all clots or blood, drilling the fragments obliquely (sparing the cartilage), and wiring them in close contact.

Fig. 23.

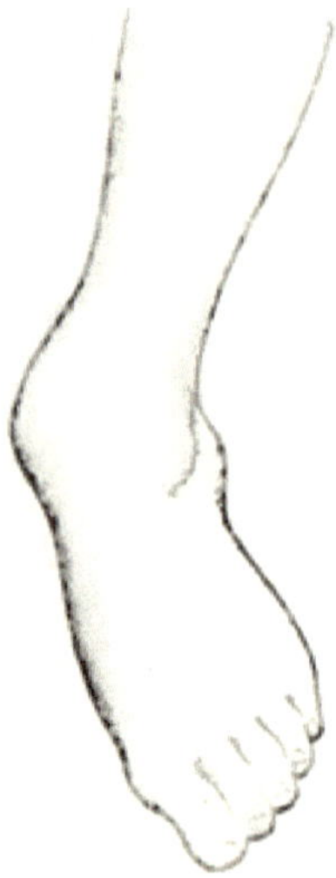

Pott's fracture.

Give the symptoms of fracture of the tibia?

Usual seat, lower third. Cause, direct or indirect violence. Deformity, slight, detected by passing the finger along the subcutaneous edge of the bone. Mobility and crepitus can usually be elicited by extension and counter-extension.

What are the symptoms of fracture of the fibula?

Cause, direct or indirect violence. Seat of fracture, lower third. Fracture of lower fifth is termed *Pott's fracture*. Symptoms obscure, disability and deformity being slight. Crepitus and mobility detected by placing the fingers over the seat of fracture and rotating, or by pressure on both sides of the suspected point.

What is Pott's fracture?

A fracture of the fibula, two to four inches above its lower extremity; the foot is displaced outward at the ankle-joint. The internal lateral ligament is frequently torn. There may be a fracture of the internal malleolus also.

What are the symptoms of Pott's fracture?

A well-marked depression at the seat of fracture. Crepitus and mobility on local pressure. The foot is twisted outwards and the sole everted by the peronei muscles; the internal malleolus projects prominently as if broken, and the fragments can be distinctly felt.

Describe fracture of both tibia and fibula.

Usual cause, indirect force. Seat of fracture, lower third. Direction of fracture, oblique. Deformity, dependent on direction of fracture, there is usually overlapping, and anterior projection of the upper or lower fragment. Diagnosis, all cardinal signs and symptoms.

How do you treat fractures of the leg?

All these fractures may be treated by the fracture-box, applying lateral compresses to correct deformity, and using extension if there is marked shortening. The fracture-box should fix the knee-joint, should be strong, and should hold the leg in such a position that the inner borders of the internal condyle, the internal malleolus, and the ball of the great toe lie nearly in the same vertical plane, and the foot is kept at right angles to the leg, pressure being taken off the heel by a pad of oakum beneath the tendo Achillis. For very marked displacement, and difficulty in retention, flex the hip and knee, lay the limb on its outer side, and bind it to a double-angled external splint for a few days, then place it in the fracture-box.

Fig. 24.

Fracture-box.

The fracture-box consists of a posterior splint, with a foot-piece and hinged sides; a pillow is placed in the box, the leg placed on the pillow, and the sides brought up and tied.

External, posterior, anterior, and straight moulded splints may also be used for these fractures.

Pott's fracture may be treated with *Dupuytren's splint*. This consists of a straight internal splint, notched at the lower end, and extending from the head of the tibia to a point four inches below the side of the foot. The upper part of the splint is fastened to the leg, a thick pad is applied to the lower portion, not extending below the internal malleolus, the foot is drawn close to the splint, in the space beneath the pad, by a figure-of-eight, so applied that there are no turns which make pressure above the external malleolus. The knee is then bent, and the leg suspended, or laid on its outer side.

Fig. 25.

Dupuytren's splint applied.

Describe fractures of the tarsal bones.

Cause, great violence.

Calcaneum or astragalus. Little displacement, unless the tuberosity is separated, when it will be drawn up by the gastrocnemius and soleus. Diagnosis depends on crepitus, pain, mobility, and great swelling.

Treatment. Fracture-box, or fixed dressing after subsidence of swelling. For separation and displacement of the tuberosity, extend the foot on an anterior or lateral splint, and flex the knee.

Describe fractures of the sternum.

Seat, about the junction of the manubrium and gladiolus.

Cause. Direct violence. Indirect violence (over flexion or extension of the body).

Deformity, readily felt. Irregularity and projection.

Crepitus and mobility by extending the body, or causing the patient to take a deep inspiration. Embarrassment of respiration, discoloration.

This injury is usually a *diastasis*, or separation of the bone at its cartilaginous junction. In this case the lower fragment projects anteriorly, the crepitus is smooth, and the true nature of the injury is suggested by its location.

Treatment. Raise the chest by placing a pillow beneath the back, force the patient to take a long breath, giving ether if necessary, and press the fragments into place.

Dressing. Broad compress, held in place by adhesive straps or bandages.

Complications. Mediastinal abscess and necrosis. Treat the former by opening at the *side* of the sternum.

If the ensiform cartilage is drawn in upon the stomach, causing distressing symptoms from pressure, it should be hooked up or resected.

Describe fractures of the ribs.

Cause. Direct or indirect violence, muscular action. Ribs commonly broken, fifth to tenth.

Ordinary seat of fracture, just anterior to the angle.

Give the symptoms of fractures of the ribs.

Crepitus and mobility, elicited by the pressure of the thumbs, passing from the sternum to the spine. Restriction of respiratory movements by a sharp pain or stitch. Displacement, if present, is *internal* from direct force, *external* from indirect.

Give the treatment for fractures of the ribs.

Adhesive strips two and one-half inches wide, running parallel to the ribs, from the spine to the sternum, and each tightly applied during expiration. The whole side of the chest is included.

If displacement exists it *must be reduced*, by pressure, by forcing the patient to inspire deeply under ether, or by hooking up with a tenaculum.

What complications accompany fractured ribs?

Laceration of the lung, pleura, or an intercostal artery.

How do you treat the complications?

Open and tie, if there are signs and symptoms of internal bleeding. Subsequent pleurisy and pneumonia are usually localized and conservative. Emphysema may require openings in the skin (strict asepsis).

In what fractures is the union ligamentous?

Neck of the femur, olecranon, acromion coracoid and coronoid processes, patella, tuberosity of the os calcis, spinous processes of the vertebræ. This is due, in part, to the difficulty in securing or maintaining apposition.

Describe fractures of the vertebræ.

Cause. Direct or indirect violence.

Seats. Spinous processes. Laminæ. Body.

Give the symptoms of fractured vertebræ.

Crepitus, mobility, and *deformity* may be detected by grasping and manipulating the spinous process, or pressing upon them, or by examination through the pharynx, in fractures of the upper cervical vertebræ. There is *immediate* paralysis of the parts below the injury, with loss of control over the bladder and rectum. Temperature of the paralyzed part is increased.

Dorso-lumbar region. Paraplegia, retention and overflow of urine, incontinence of fæces.

Dorsal region. Second to eleventh dorsal. Paralysis of abdominal muscles, and muscular coat of intestines. *Expiration* markedly embarrassed from involvement of serratus posticus inferior, quadratus lumborum, sacro-lumbalis, longissimus dorsi.

Cervico-dorsal, cervical. If above the fifth and sixth cervical vertebræ, paralysis of the arms, and more marked embarrassment of respiration from involvement of the long thoracic nerves (fifth and sixth). If above the third and fourth vertebræ, instant death, from involvement of the phrenic. Fractures of the atlas and axis need not be immediately fatal, since the canal is so roomy that the cord may not be encroached upon.

Odontoid process will cause a prominence in pharynx from subluxation of the axis. Rigid maintenance of head in one position.

How do you treat fractures of the vertebræ?

If there is displacement, reduce by extension and manipulation. Place the patient on an air or water bed, guarding against bedsores by frequent washings with whiskey and alum, and careful padding with soft pillows. Move the bowels by enemata. Draw the water regularly with a soft, thoroughly aseptic catheter. In fractures about the neck, support by means of short sand-bags.

How do you treat fractures of the extremities complicated by delirium tremens?

Carefully pad with raw cotton, and put on a fixed dressing, as plaster or silica; when dry, bind the limb in a soft pillow.

LUXATIONS OR DISLOCATIONS.

Define luxation.
A luxation is the displacement of the articular surfaces of a joint from their normal relation to each other.

Name and define the various kinds of luxation.
In regard to cause—

1. Traumatic, due to sudden force.
2. Pathological or spontaneous, due either to alterations of the joint from disease (coxalgia), or to paralysis of the surrounding muscles.
3. Congenital, due to congenital malformation of the joint (luxation produced by violence in delivery is not congenital).

Further, we have luxation classed as—

Complete. An entire separation of the articular surfaces from each other.

Partial (subluxation). The articular surfaces remain in contact through a portion of their surface.

Recent. When sufficient time has not elapsed for inflammatory changes seriously to impede reduction.

Old. When such changes have taken place.

Simple, *compound*, and *complicated* are applied to luxations precisely as in case of fracture.

What are the causes of luxation?
(1.) PREDISPOSING.—1. The nature of the joint (ball-and-socket joint). 2. The position of the joint. 3. The condition of the surrounding soft parts. (Paralysis, relaxation, and previous inflammation.) 4. Age and sex of the patient. (Adult male.)

(2.) EXCITING.—Direct or indirect violence. Muscular force.

What are the cardinal symptoms of luxation?
1. Change in the shape of the joint.
2. Alteration of the normal anatomical relations of the bony prominences about the joint, the displaced bone being often felt in its abnormal position.
3. Alteration in the length of the limb.

4. Rigidity, or restricted motion of the affected joint.
5. Alteration in the direction of the axis of the bone.

In addition we have the symptoms attendant on all traumatisms.

Pain of a dull sickening character.

Swelling often very great.

Discoloration diffused about the joint.

How do you distinguish luxations from fractures?

1. In luxation there is no harsh crepitus.
2. There is rigidity in place of undue mobility.
3. The deformity, when reduced, has not the same tendency immediately to recur.

The pain is not so intense, the swelling and discoloration not so rapid, and at times the smooth displaced articular surface may be felt, while in fracture, except epiphyseal, the surfaces would necessarily be rough.

What articular changes take place in luxation?

Rupture of capsular ligament, with stretching or tearing of surrounding vessels, tendons, muscles, and nerves.

Prompt reduction of the bone favors the repair of the injury. If the bone is not reduced the articular cavity becomes filled up, the prominences rounded off; a new socket is formed about the displaced head of the bone. The surrounding soft parts become shortened and atrophied, and adhesions between the bone and the vessels or nerves often take place.

What is the prognosis in luxation?

Usually a weakened joint. If the dislocation is not reduced, permanent disability, which, however, is rarely absolute.

How do you treat luxation?

Reduce by either *manipulation* or *extension*.

Describe the methods of reduction.

1. *Manipulation* consists in so placing and moving the parts that muscles and ligaments are relaxed, articular prominences are disentangled from each other, and the head of the bone is either

drawn by the muscles, or pushed by moderate force into its proper position.

2. *Extension* consists in overcoming resistance by force— this force may be applied by the hands, by wet sheets or bandages fastened about the parts, or by multiplying pulleys. When the tension is sufficient to overcome all resistance the bone is pushed into its proper position. Retain in position by splints and bandages.

How do you treat the inflammatory symptoms?

Treat by evaporating lotions or counter-irritants. The diet should be restricted and the bowels kept opened.

How do you prevent anchylosis?

By passive motion, beginning in seven to ten days, or as soon as inflammatory symptoms subside.

What complications attend luxations?

1. *Fracture.* Treat by setting and splinting the fracture, then reducing the luxation.

2. *Rupture of a large artery*, indicated by a rapidly increasing, fluctuating, pulsating swelling. Treat by rest and pressure, or ligate both ends at the point of injury, if it can be found. If this is impossible, make a formal ligation of the artery above.

3. *Injury to nerve-trunks.* Treat by friction, electricity, massage, incision and suture.

4. *External wound, or compound luxation.* If no extensive injury to the joint, thoroughly disinfect, replace, close the wound, and fix. If the bone is comminuted, resect.

How do you treat an old luxation?

Loosen adhesions and relax contracted muscles and ligaments by passive motion. Endeavor to replace the bone by manipulation; that failing, use force.

What accidents may occur in the reduction of old luxations?

Fractures. Set at once, and give up further attempt.

Rupture of important muscles. Put at rest.

Rupture of principal artery. Ligation of artery above, or ligation of both ends at point of rupture, or amputation.

Ruptured vein. Pressure.

If an old luxation gives little pain on movement let it alone, as the prognosis is good. If great pain, try to reduce, since the pain will prevent the patient from endeavoring to restore function.

Special Luxations.

Describe luxations of the lower jaw.

Direction is forward. May be *unilateral*; more commonly *bilateral*. May be *partial*, the condyles resting on the articular eminence, or *complete* the condyles slipping into the zygomatic fossa.

Cause. Violence or muscular force, applied when the mouth is widely opened. In this position the condyles ride well up on the articular eminence, and may be easily pulled forward by the action of the external pterygoid, and masseter, or by direct force. This displacement may occur in yawning, laughing, etc.

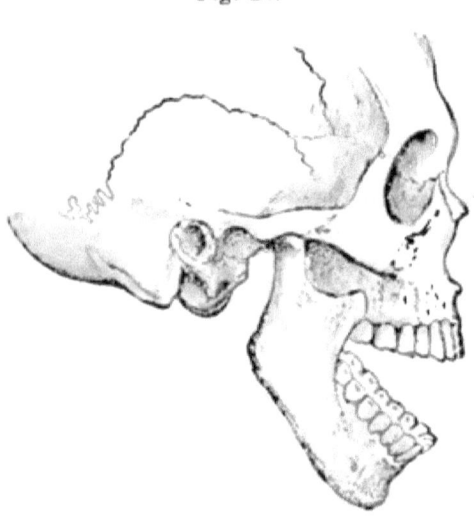

Fig. 26.

Complete luxation of the lower jaw.

Give the symptoms of dislocation of the jaw.

Bilateral. Mouth *widely opened* and *rigid*, lower jaw thrust forwards, lips cannot be approximated, hence dribbling of saliva. A *depression* is felt in the normal position of the condyle, the latter forming a *prominence* in front. Difficult deglutition, pain, and swelling.

Subluxation. Condyles and lower jaw slightly anterior to normal position, jaw rigidly closed, and great pain.

Unilateral luxation. Mouth less widely opened, lower jaw projected anteriorly, and thrust towards sound side; displaced condyle detected on the affected side; other symptoms as in bilateral luxation.

Give the treatment for dislocations of the inferior maxilla.

Disengage the head of the bone from the zygomatic fossa, when the internal pterygoids and the masseter and temporal muscles will pull it in place. This can be effected by pressing downward upon the molar teeth of the lower jaw, at the same time pulling up the chin. The protected thumbs of the surgeon's hand are placed upon the molar teeth, exerting force downward and backward, while, with the fingers, the chin is pressed up; or wedges may be inserted between the molar teeth of the lower and upper jaws on each side, and the chin forced directly upwards.

Unilateral luxation. Force exerted as before, on the affected side of the jaw.

Sub-luxation. Slip a case-knife between the teeth of the upper and lower jaws, and pry them open, when the muscles promptly reduce the displacement.

Describe luxation of the ribs.

Occurs at *costo-chondral* or *vertebral articulations*. If at vertebral extremity, usually associated with fracture.

Symptoms as in fracture, except no crepitus.

Treatment as for fracture.

Describe luxation of the vertebræ.

Nearly always complicated by fracture.

Usual seat. Cervical region.

Symptoms. Sudden *paralysis*, *rotary* or *angular deformity*, and *rigidity*.

Treatment. Reduce by extension and counter-extension in the line of the body. Treat subsequently on a water-bed as for fracture.

Describe luxations of the clavicle.

More frequent at acromial than at sternal extremity.

Sternal extremity. Forward, by force applied to front of

shoulder. Most common. *Backward*, by force applied to back of shoulder or applied directly on sternal extremity of bone. *Upward*, very rare, by force applied to shoulder from above.

Give the symptoms of luxations of the sternal end of the clavicle.

Shoulder falls towards median line, *pain* on motion. *Smooth articulating surface* of bone felt in its *abnormal* position leaving a depression in the seat of its articulation. If luxation backwards or upwards there may be dyspnœa, dysphagia, or venous congestion of head, from pressure.

Give the treatment for luxation of the sternal end of the clavicle.

Forward and *backward luxations*. Reduce by knee between scapulæ, pulling shoulders back, and pressing the bone in place.

Upward luxation. Reduce as above, or by placing a large pad in the axilla, pressing the humerus to the side, and pushing the bone in place.

Dressing. Forward luxation. Flex arm and apply a Velpeau or Désault, keeping the displaced bone in place by compress and adhesive strips.

Backward. Posterior figure-of-eight and Velpeau or Désault.

Upward. Velpeau bandage, with compress and adhesive strips if persistent deformity.

Describe luxations of the acromial extremity of the clavicle.

Really luxations of the scapula.

Direction upward, rarely downward below acromion, or still more rarely, below coracoid process.

Cause. Direct blow on scapula.

Give the symptoms of luxation of the acromial end of the clavicle.

Upward luxation. Shoulder falls down and in. Arm cannot be raised over head. *Outer extremity of clavicle very prominent*, overriding acromion process.

Downward luxation. Same symptoms, except the acromion is prominent; the clavicle leads down to the axilla and can be felt in its abnormal position beneath the acromion or coracoid process.

Give the treatment for luxations of the acromial end of the clavicle.

Reduce, by pulling the shoulder backwards and pressing the bone in place. Place a compress over the acromial extremity of the clavicle and fasten it in place by broad straps passing over it and around the point of the elbow. Apply a Velpeau bandage. In all luxations of the clavicle *reduction* easy, *retention* difficult.

Keep up the dressing for five to six weeks, then carry the arm in a sling for some time.

Describe dislocation of the scapula.

By this is meant the slipping out of the inferior angle of the bone from beneath the latissimus dorsi.

Cause. Paralysis of the serratus magnus, or violence.

Symptoms. Wing-like projection, pain, and weakness of shoulder.

Treatment. Broad belt which will keep the inferior angle of the scapula close to the chest.

Describe the shoulder-joint.

Characterized by a large ball and small socket, allowing great freedom of motion.

Ligaments. 1. *Capsular.* Very lax, weakest at lower part, attached to margins of glenoid cavity and to anatomical neck of humerus.

2. *Coraco-humeral.* Passing from root of coracoid process downward and outward to the front of the great tuberosity.

3. *Glenoid.* A triangular ring of fibro-cartilage, deepening the glenoid cavity. The joint is further strengthened by the tendon of the biceps passing directly over it, and invested in a prolongation of its synovial membrane.

Name the luxations of the shoulder-joint.

Four in number. *Subglenoid, subcoracoid, subclavicular,* and *subspinous.*

144 ESSENTIALS OF SURGERY.

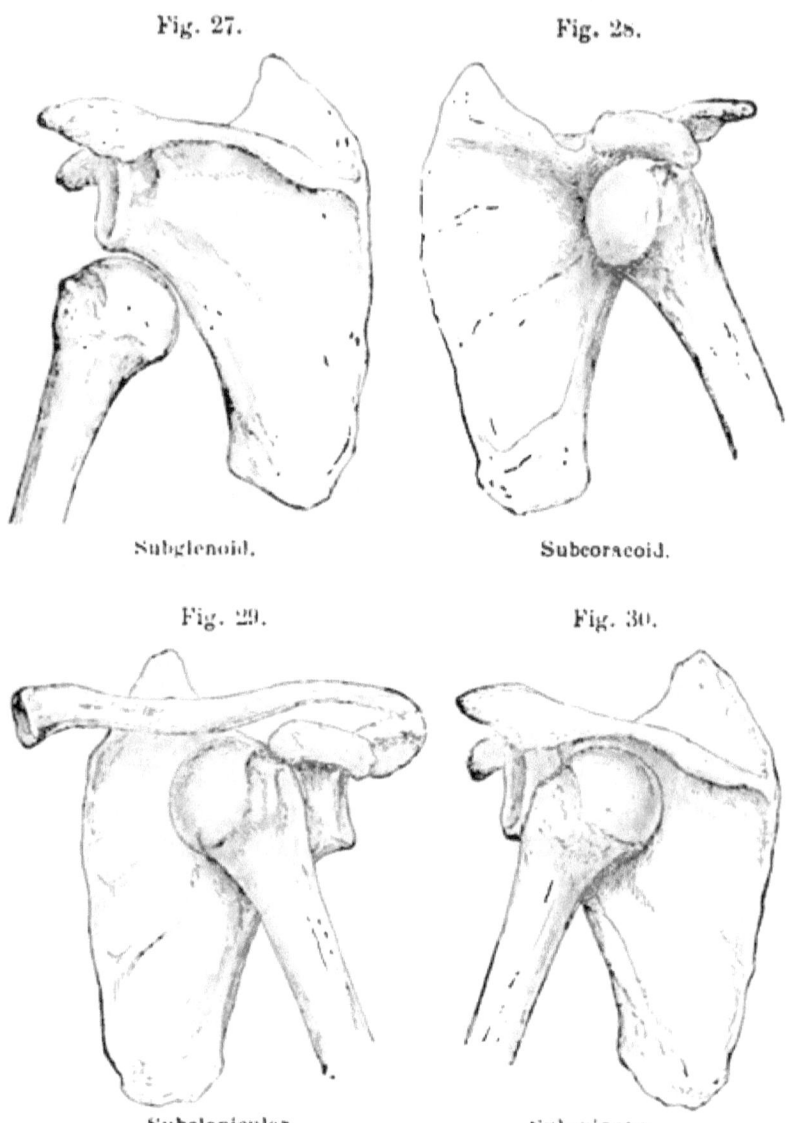

Fig. 27. Subglenoid.
Fig. 28. Subcoracoid.
Fig. 29. Subclavicular.
Fig. 30. Subspinous.

What symptoms are common to all shoulder luxations?

1. Flattening and squareness of the shoulder, with apparent projection of acromion process.

2. A depression beneath the acromion process, where the head should lie.

3. The head of the bone can be felt in its abnormal position.

4. The vertical measurement of the shoulders, from the axilla around the acromion process, is one or two inches greater on the affected side than on the sound side.

5. With the elbow brought close to the body, the patient cannot place the hand of the injured side upon the opposite shoulder (except in subspinous).

6. Alteration in the axis of the humerus.

7. Rigidity, pain, swelling, discoloration, etc.

What symptoms characterize subcoracoid luxation?

This is the most common luxation. 1. Head of bone can be felt in the upper and anterior part of the axilla, beneath the coracoid process.

2. The humerus stands from the side (deltoid), and is somewhat oblique in direction, the elbow being carried back (latissimus dorsi and teres major).

3. Pressure on axillary plexus especially marked, and consequent numbness and tingling in the arm and forearm.

What symptoms characterize subglenoid luxation?

Next in frequency. Head of bone rests on axillary border of scapula, and can be felt in the axilla. Elbow carried far from the side (deltoid). *Lengthening of the arm*, measured from the acromion process to the external condyle of humerus.

What symptoms characterize subspinous luxation?

Elbow carried somewhat forward (pect. major), and bone rotated inward (subscapularis), the forearm being thrown across the chest. *Head* of bone felt on dorsum of the scapula. *Coracoid process prominent*.

What symptoms characterize sub-clavicular luxations?

Head of bone seen or felt internal to coracoid process, and below clavicle, much *laceration* of muscles attached to tuberosities.

Elbow out and back. All the characteristic symptoms.

How do you treat luxations of the humerus?

1. *Reduce by manipulation.*

Subglenoid, subcoracoid, and *subclavicular.* Flex forearm on arm (relax long head of biceps); raise the arm from the body (relax deltoid and supra-spinatus); rotate the humerus outward (relax infra-spinatus and teres minor); make forcible traction upon the humerus with one hand, sweeping it to the side of the body and rotating it inward, carrying the forearm across the chest, while with the other hand in the axilla the head of the bone is pressed into place.

Subspinous. Flex the forearm, grasping the elbow, carry the humerus from the side, rotate inward (subspinous), and with the thumb press the head of the bone in place.

2. *Reduce by extension.*

Heel in the axilla. Patient supine, surgeon sits down beside him, places his heel (unbooted) in the axilla, and makes traction

Fig. 31.

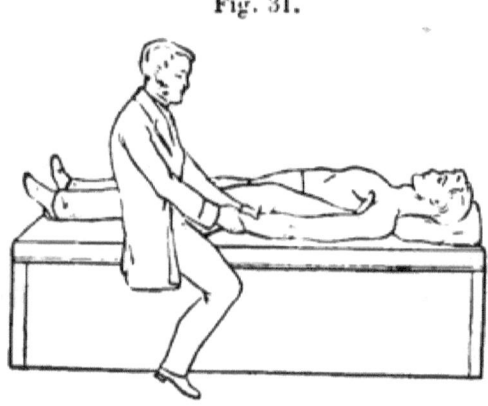

Reduction by extension.

on the wrist, at first directly downwards. If the luxation is not reduced, the humerus is carried across the chest by pulleys. Rarely employed except in old dislocations.

After treatment. Arm to side and axillary pad for a week, passive motion for two weeks, then allow patient to use arm.

Old luxations. If more than three months have elapsed and there is a fair amount of motion, do not attempt to reduce.

Luxations of Elbow.

What dislocations may occur at the elbow-joint?
Radius. Forwards, backwards, outwards.
Ulna. Backwards.
Both bones. Forwards, backwards, inwards, outwards.
Ordinary luxation. Both bones backwards.

Describe backward luxation of both bones.
Cause. Fall on palm of hand.
May be *complete*, when coronoid process of ulna is lodged in olecranon fossa of humerus, or *incomplete*, when coronoid process rests upon the articulating surface of the humerus (trochlear).

Give the symptoms of backward luxations of the radius and ulna.
1. Olecranon projects posteriorly, is out of line with condyles, and the distance between it and the condyles is greatly increased. Head of radius felt behind external condyle.
2. A smooth, broad, rounded projection, the articular extremity of the humerus, can be felt in front of the elbow, below the joint crease.
3. The forearm is flexed, supinated, and rigid.
4. Shortening, from external condyle to styloid process of radius.

Give the symptoms of forward and lateral luxations of radius and ulna at the elbow.
Both bones forward, very rare, forearm lengthened, condyles of humerus prominent, sigmoid notch can be felt in front of arm.
Lateral luxation of both bones. Great deformity. The articulating extremity of the radius or ulna can be felt in their abnormal positions, with marked projection of the condyle from which the bones are displaced; joint widened, forearm flexed and pronated.

Give the symptoms of luxation of the ulna at the elbow.
Direction, always backward. The symptoms are the same as for both bones backward, except that the head of the radius

can be felt in its normal position, and the forearm is *shortened only on its ulnar aspect.*

Give the symptoms of luxations of the radius at the elbow.

Directions, forward, backward, outward.

Forward, due to force applied in supination.

Backward, due to forcible pronation. In both, the head of the bone can be felt in its abnormal position, leaving a hollow below the capitellum of the humerus. Motion restricted.

Give the treatment for luxations at the elbow.

Dislocation of ulna or of both bones.

Forcible flexion of forearm over the knee placed in the bend of the elbow; or forcible extension of the forearm, followed by flexion.

Radius. Anterior luxation. Flexion of forearm, direct pressure upon head of radius, and forced pronations.

Posterior luxation. Flexion of forearm, forced supination, direct pressure.

Dressing. Anterior angular splint one week, with compress, in case of radius; passive motion daily. These luxations become old in one or two weeks. If attempt to reduce an old luxation is made, first break up adhesions.

Describe luxations of the carpal extremity of the ulna.

Cause. Forward, violent supinations. *Backward*, violent pronations.

Symptoms. Projection, with ordinary symptoms. Triangular cartilage always broken.

Treatment. Press bone in place, apply compress and bandage, or adhesive plaster, keep up support for several months.

Describe luxations of the carpus.

The wrist-joint is formed by the radius and triangular cartilage articulating with scaphoid, semilunar, and cuneiform bones.

Cause of luxation. Force applied to hand in front or behind.

Direction. Backward or forward.

Symptoms. Thickness of wrist greatly increased. Distance between styloid process of radius and base of metacarpal bone of thumb lessened. The smooth round projection of the carpal bones felt on one surface of the wrist, the more irregular projection of the lower extremity of the radius felt on the opposite surface. Rigidity, pain, etc. Hand somewhat flexed in posterior luxation, somewhat extended in anterior luxation.

Treatment. Posterior displacement. Flex, press carpus forward, on first sign of slipping into place suddenly extend.

Anterior displacement. Extend, press carpus backward, and on first sign of slipping into place suddenly flex. Reduction may be effected by extension and counter-extension.

Splint and begin passive motion as soon as inflammation subsides.

Describe luxation of the individual carpal bones.

Direction. Backwards.

Cause. Direct force.

Common seat. Os magnum.

Symptoms. Projection at base of third metacarpal bone, with ordinary symptoms of luxation.

Treatment. Extend, press into place, and apply palmar splint with compress.

What luxations may occur in the hand?

Metacarpus. Rare.

Direction. Backwards.

Symptoms. Prominence and shortening.

Treatment. Extension, pressure, and palmar splint.

Phalanges. Seat. Usually first phalanx of index or little finger. Direction. Anterior or posterior.

Symptoms. Shortening and undue prominence.

Treatment. Traction, or extreme extension and forcing bone into place by direct pressure.

What is the most difficult luxation to reduce?

Backward displacement of first phalanx from the metacarpal bone of the thumb.

What is the cause of difficulty?

The head of the metacarpal bone slips in between the two heads of the short flexor of the thumb, and is embraced the more tightly, in proportion to the amount of traction exerted on the displaced phalanx.

What are the symptoms of backward luxation of the first phalanx of the thumb?

Head of metacarpal bone felt in palmar aspect of hand. Proximal phalanx extended, terminal flexed. Immobility, etc.

Give the treatment.

Forcibly adduct the metacarpal bone into the palm, extend the phalanx far backward till the thumb-nail nearly touches the wrist, then suddenly flex on the metacarpal bone, at the same time pressing the displaced phalanx into position. If this method fails, tenotomy of the flexor brevis pollicis.

Name the ligaments of the hip-joint.

1. *Cotyloid*, a rim of fibro-cartilage deepening the acetabulum.
2. *Transverse*, bridges over the notch, and is continuous at each end with

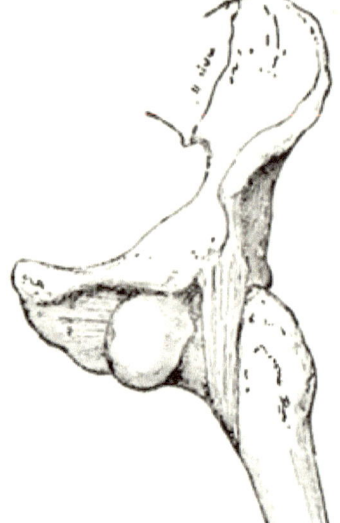

Fig. 32. Fig. 33.

Y-ligament. Obturator luxation.

3. *Ligamentum teres*, which passes to a depression in the head of the femur.

4. *Capsular*, encircling the acetabulum above and attached to anterior intertrochanteric line, to inner and upper border of the great trochanter, and posteriorly and below to the junctions of the middle and outer thirds of the neck of the femur.

5. Y-*ligament*, a thickened part of the capsular ligament rising from the anterior inferior iliac spine and splitting as it passes down to be inserted into the intertrochanteric line. Lower and inner part of joint is weakest.

Name the dislocations of the hip-joint.
1. Up and back on *dorsum ilii*. Iliac.
2. Back in *sciatic notch*. Ischiatic.
3. Forward and down in *obturator foramen*. Obturator.
4. Forward and up on pubis. Suprapubic.

Causes. Force applied when the limb is abducted.

What symptoms characterize the backward luxations?
1. *Dorsum ilii.* Upwards and backwards. *Bulging of hip* from displaced trochanter major, which *lies above Nélaton's line* and nearer the anterior superior spinous process of the ilium than on the sound side.

Shortening, one and one-half inches. Pressing the fingers into the groin over the femoral vessels, their firm base or support is gone, *a hollow is felt* instead. *Head* of the bone may be felt beneath glutei muscles.

Position of leg. Adduction and inversion due to Y-ligament. *Knee* rests against *lower third* of opposite thigh. *Great toe* rests on *instep* of opposite foot.

Rigidity, pain, swelling, etc.

2. *Ischiatic or sciatic luxation* (below the tendon of the obturator).

Symptoms the same, but less marked. Less shortening, adduction, and inversion.

Knee touches, but does not cross opposite knee. *Ball of great toe* rests on *metatarsal bone* of opposite side.

Fig. 34.

Dorsum illi.

Fig. 35.

Ischiatic.

Give the treatment of backward luxations.

Manipulation. Flex leg on thigh (relax hamstring muscles), thigh on abdomen, and still further adduct to relax anterior part of capsule; then maintaining flexion, circumduct (abduct and rotate) outward as far as possble, bringing the leg suddenly down to an extended position by the side of its fellow. By this means the head of the bone is made to retrace the steps by which it escaped, and is wound in place by the Y-ligament.

Manipulation failing, try

Extension. Secure *counter-extension* by strapping the pelvis to the floor or bed. Make *extension* by flexing the thigh on the pelvis and pulling directly upward.

Fig. 36.

Manipulation for reduction of backward luxation.

Give the symptoms characterizing forward luxations.

Obturator luxation forward and downward.

1. Psoas, iliacus, external rotators, and Y-ligament put upon the stretch, hence *Eversion* and *abduction* with *slight flexion*, thigh being carried somewhat forward.

2. Flattening of hip and, possibly, detection of bone in abnormal position.

3. Slight lengthening (one-half inch).

4. *Fixation*, swelling, and other signs.

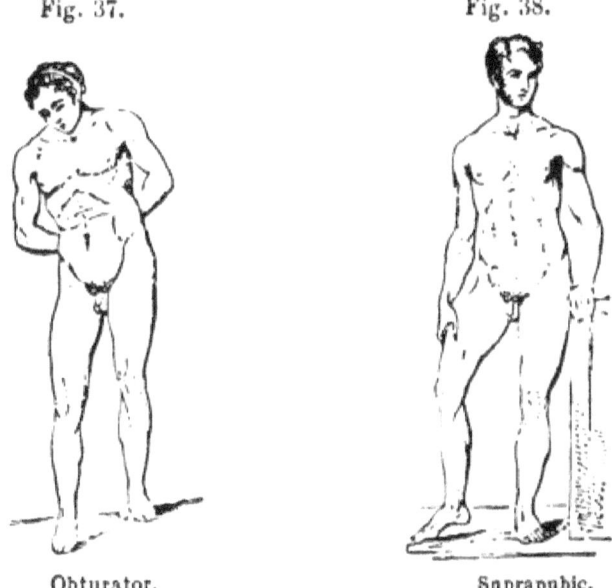

Fig. 37. Fig. 38.

Obturator. Suprapubic.

Suprapubic luxation. 1. Head of bone readily felt on pubis, to outer side of femoral artery.

2. Shortening (1½ inch), with *very marked eversion* of foot and knee, heel inclining towards opposite one.

3. *Trochanter* may be *internal* to anterior superior spinous process.

4. Depression over acetabulum.

Give the treatment of forward luxations.

Reduction. *Obturator*—Flex leg on thigh, thigh on abdomen, abduct somewhat, then *circumduct inward*, carrying thigh over body and making *internal rotation*, and bring the leg down to the side of its fellow.

Suprapubic as for obturator, but do not carry the thigh so far across the body.

Fig. 39.

Manipulation for reduction of forward luxations.

Give the after-treatment of all luxations at the hip-joint.

The knees bandaged together (a towel between them) for ten days, passive motion in bed for two weeks, wearing moulded support for three months.

Name the internal ligaments of the knee-joint.

1. *Anterior and posterior crucial.*
2. *The transverse ligament,* binding together the two semilunar cartilages.
3. *The coronary ligament,* connecting the outer borders of the semilunar cartilages to the head of the tibia.
4. *Ligamentum mucosum,* a process of synovial membrane, and *ligamenta alaria,* its fringed borders.

Describe luxations of the knee-joint.

Cause—great violence. *Directions*—Forward, backward, inward, and outward.

Lateral dislocations mostly *incomplete;* more common than antero-posterior.

Give the symptoms of backward and forward luxations of the knee-joint.

1. *Shortening.* 2. *Great deformity.* The articulating extremities of the femur and tibia being readily felt in their abnormal positions.

Give the symptoms of lateral luxations of the knee-joint.

No shortening, but marked lateral projection of the tibia, with a depression above; condyle of femur prominent on opposite side, with a corresponding depression below.

Give the treatment of luxations of the knee-joint.
Treatment. Flex the thigh, make extension, and push bone in place. *Reduction easy.*

Apply a posterior straight splint. Treat the synovitis (cold, counter-irritation, etc.), and begin passive motion as soon as acute inflammatory symptoms subside. A knee-cap must be worn when the patient is allowed to walk.

In what directions may the patella be dislocated?
1. *Outwards.* (Most common, from oblique attachment of quadriceps tendon.)
2. *Inwards.*
3. *Quarter rotation.*
4. *Half rotation.*

Give the symptoms of luxation of the patella.
Outward and inward luxations.
1. Knee flattened and broadened.
2. *Sulcus* in normal position of patella.
3. *Patella* readily found in abnormal position.

Give the treatment for lateral luxations of the patella.
Anesthetize, flex thigh or abdomen, extend leg on thigh, forcibly depress the margin of the patella furthest from the centre of the joint, when its inner edge being raised and freed, will be snapped into place by the quadriceps.

Give the symptoms of rotatory luxation of the patella.
Quarter rotation. 1. Sharp edge of patella felt prominently under skin. 2. Leg fixed in extension.

Half rotation. 1. Tendo patella stands rigidly out and is twisted. 2. Smooth articular facets of under portion of patella felt. 2. Limb rigidly extended.

Treatment. Anesthetize. Rapid flexion and extension of leg on thigh. If this fails, employ direct pressure.

Describe luxation of the semilunar fibro-cartilage.
Causes. Twists of foot or leg while the knee is flexed.

Directions. Inward towards spine of tibia, outward.

Give the symptoms of luxation of the semilunar cartilage.

1. If outward, a *projection may be felt* between tibia and condyle of femur. If inward, a *depression may be noted* in the same position.
2. *Sudden, violent, sickening pain.*
3. *Leg fixed in semi-flexion.*
4. *Rapid effusion into joint.*

Give the treatment for luxations of the semilunar cartilage.

Forcible flexion, straight posterior splint. Treat accompanying synovitis. A knee cap must subsequently be worn.

Describe luxations at the ankle-joint.

Directions. Outwards, inwards, forwards, backwards, upwards (between tibia and fibula).

May be *complete* or *incomplete*. *Complications.* Frequently *fractures*.

Outward. Always accompanied by fracture of fibula, frequently of internal malleolus also, or rupture of internal lateral ligament.

Symptoms. As in Pott's fracture (p. 32). Foot everted. Internal malleolus prominent.

Inward. Rare. Accompanied by fracture of tibia.

Symptoms. 1. Foot inverted. 2. External malleolus prominent and nearly touching ground. 3. Depression over seat of fracture.

Backward. 1. Marked shortening of foot with toes pointed downward. 2. Lengthening of heel.

Forward. 1. Lengthening of foot. 2. Heel less prominent. 3. Tibia lies close to tendo Achillis, which is relaxed.

Upward. Caused by heavy fall on feet.

Symptoms. Joint very wide, malleoli may be prominent and nearly on a level with the sole.

Give the treatment for luxations of the ankle-joint.

Reduce. Flex leg on thigh, extend ankle-joint to relax muscles of calf. *Extension* must be made at the foot. *Counter-extension* at the thigh, while by manipulation and pressure the bones are replaced in their proper position.

After treatment. Control inflammatory symptoms by evaporating lotions. Fracture-box, or moulded splints for two weeks, then passive motion.

Describe luxations of the astragalus.
Directions. Forward, backward, outward, inward.
Forward, most common.
Cause. Violent twists.
Symptoms. In all these luxations the malleoli are nearer the sole than they should be.
Forward. A round smooth swelling upon the instep, with ordinary signs of luxation.
Backward. 1. Hard prominence between tendo Achillis and malleoli. 2. End of tibia and fibula prominent anteriorly. 3. Foot apparently shortened.
Lateral luxations. If astragalus is thrust outward the foot is displaced inward. Internal malleolus very prominent.
Inward luxation. Foot displaced outward. External malleolus prominent.
Reduce. By traction and direct pressure, under ether. Failing, perform tenotomy, dividing all resisting structures. If skin sloughs over projecting astragalus, remove the bone.
Failing to reduce, put in fracture-box and treat as ankle luxation.

Give the differential diagnosis between fracture of the surgical neck of the humerus, and luxation about the shoulder-joint.
In fracture, *crepitus, unnatural mobility. Head of the bone in its normal position, but not moving with shaft.* Deformity readily overcome, but at once recurring on removal of reducing force; acromion not especially prominent, and no undue space beneath it; jagged bone ends may be felt; very acute pain. *Arm hangs to the side.*

Luxation. *No crepitus. Rigidity.* A hollow in the normal position of the head of the bone. Detection of head of bone in abnormal position, *moving with the shaft.* Deformity reduced with difficulty, after reduction the bone remains in its normal position; acromion prominent, with a space beneath. *Shoulder flattened and squared. Arm stands from the side.*

Give the differential diagnosis between supracondyloid fracture of the humerus, and backward luxation of the radius and ulna.

Fracture. Crepitus, mobility, and all cardinal signs; olecranon and internal and external condyle in their normal relation to each other; no shortening from external condyle to styloid process of radius, shortening from acromion to external condyle.

Luxation. Immobility, and all the signs of luxation; olecranon displaced backward from its normal position in relation to internal and external condyles; shortening from external condyle to styloid process of radius, no shortening from acromion to external condyle.

The differential diagnosis between any fracture, and a luxation in the same region, may readily be given by bearing in mind the cardinal symptoms of each affection.

Sprains.

What is a sprain?

The twisting of a joint, by which the soft parts about it are stretched or torn. Muscles, tendons, ligaments, nerves, and bloodvessels may be involved.

What is a sprain fracture?

The tearing away of scales of bone to which ligaments are attached.

What are the symptoms of sprain?

Pain and swelling due to both extravasation of blood, and inflammatory effusion within and without the joint. *Discoloration and loss of function.*

Give the treatment of sprain.

Hot fomentations, or hot bath, lasting for several hours, followed by *pressure bandage* for two to four days. Passive motion and massage as soon as the inflammatory symptoms begin to subside. Or, cold applications and evaporating lotions, followed by pressure and massage.

Describe sprains of the back.

Symptoms. Pain, stiffness, and disability, appearing some time after the injury. There may be apparent paresis, together with retention of urine and fæces, due to the pain caused by motion. There is sometimes hæmaturia.

Treatment. Rest in the most comfortable position for a few days, with local depletion (leeches), hot moist applications (antiseptic poultices), and counter-irritants. Then massage and use. If there is great pain on motion, a plaster bandage may be applied, to be removed as soon as possible.

Wounds of Joints.

What symptoms characterize joint wounds?

Symptoms of acute inflammation, with distension, due to effused blood and synovial fluid, and escape of the latter through the external wound.

If the contents of the joint cavity become infected, the characteristic symptoms of an acute suppurative synovitis and arthritis will appear, together with the high fever (103°–105°), and marked constitutional symptoms of the affection.

How do you treat a wounded joint?

If uncertain as to whether the joint is wounded, *do not probe*, but treat as a wounded joint.

1. *Small incised wounds.* Thoroughly disinfect the wound area, close promptly, using sutures if necessary. Cover with a scale of iodoform and collodion. Carefully splint in the easiest position, and apply cold by means of ice-bags. If marked local and general inflammatory symptoms appear, open the joint, and treat as—

2. *Large or lacerated wounds.* Thoroughly disinfect the entire wound area. Wash out the synovial cavity with 1 : 1000 bichloride solution, finishing with 1 : 5000. Make a counter opening, and insert drainage-tubes. Suture the external wound, apply an antiseptic dressing, splint most carefully, and elevate the limb.

Synovitis.

What is synovitis?

An inflammation of the synovial membrane of a joint. It may be *acute* or *chronic*. There may be an effusion consisting of synovia and serum, constituting *serous synovitis*. This effusion may become infected, causing *purulent synovitis*.

What are the causes of synovitis?

Exposure to heat or cold, traumatism, rheumatism, gout, syphilis, tuberculosis, gonorrhœa, and pyæmia.

Give the symptoms of acute synovitis.

Pain, intense, bursting. Worse at night.

Tenderness. Slightest touch or motion unbearable.

Swelling. Fluctuates, takes the shape of the synovial sac, and appears at certain portions of the joint. (At the sides of the quadriceps tendon and beneath the patella, in the knee-joint; at the sides of the olecranon and triceps in the elbow-joint.)

Muscular atrophy. Inflammatory fever, with local *heat* and *redness.*

If *suppuration* ensues, these symptoms, both local and constitutional, are aggravated; the patient has chills, the fever shortly becomes typhoid in type, and the joint becomes *red* and *œdematous.*

How do you treat acute synovitis?

Carefully splint in the position which will leave the most useful limb should anchylosis occur. (Elbow at right angles, knee straight.) Leeches and an ice-bag in the early stages. *Aspirate* if the synovial sac becomes greatly distended. Light diet, opium to relieve pain, regulate the bowels.

If *suppuration ensues*, incise, irrigate, drain, and dress antiseptically. Stimulants, tonics, and generous diet.

Describe chronic synovitis.

May result from acute. Synovial membrane may become *thickened* and *indurated* from venous congestion, or pass into a state of *fatty* or "*pulpy*" *degeneration*. *Fluid in the synovial sac* usually *considerable* in amount; clear, or slightly opalescent.

Muscular atrophy commonly present. Symptoms of inflammation slight or wanting. *Disability* not absolute, joint weak, but can be used.

Give treatment of chronic synovitis.

Counter-irritation by blisters, or tr. iodin. *Pressure* by elastic bandage. *Unguent. hydrarg. cum belladon.* locally. *Fixation* by means of plaster bandages. *Injections* of tr. iodin, and distilled water, equal parts of each, into the joint. *Treatment of associated systemic conditions*, as rheumatism or syphilis.

Describe hydrarthrosis.

Hydrarthrosis or *hydrops articuli* is a serous effusion into a joint. It may arise from acute or chronic synovitis, or spontaneously.

Symptoms and treatment as for chronic synovitis. *Open* and *drain* if everything else fails.

Arthritis.

What is arthritis?

Arthritis is an inflammation beginning in either the synovial membrane or the bone, and affecting all the structures of a joint.

What are the varieties of arthritis?

Acute. Chronic. Traumatic and infective (pyæmia, gonorrhœa, etc.), usually acute. Diathetic (struma, gout, rheumatism), frequently chronic.

What are the symptoms of acute arthritis?

Pain. Throbbing, tensile, *worse at night*. The limb is subject to *spasmodic startings* during sleep, which, from the pain they provoke, will cause the patient to wake suddenly with a cry ("osteocopic cry").

Tenderness. Developed to its most extreme extent.

Swelling. Involves the entire joint area.

Crepitus. May be felt when the cartilages are eroded.

Preternatural mobility. Although the joint is rigidly fixed by the muscles, examination under either will show softening and relaxation of ligaments, and the possibility of producing motions not normal to the joint.

Atrophy. Muscles of the affected limb rapidly waste.

Heat, redness, and œdema. Especially when pus is formed.

Fever. Ranges high, accompanied by rigors when there is suppuration, and quickly passes to the typhoid or the hectic type.

What symptoms distinguish arthritis from synovitis?

In arthritis. Starting pains at night. Swelling more diffused about the joint and doughy rather than fluctuating. Crepitus. Unnatural mobility and atrophy more marked. Constitutional symptoms more serious.

Give the treatment for acute arthritis.

Absolute rest in a favorable position (splint), with elevation, and the application of cold or heat.

If suppuration ensues, open freely, drain thoroughly, and treat antiseptically.

In some cases of traumatic arthritis, or arthritis secondary to acute epiphysitis, amputation may be necessary, if the patient steadily fails after opening and draining.

Constitutional treatment. Stimulants, tonics, and generous diet.

What is the usual cause of acute arthritis in infants?

An acute epiphysitis which suppurates, and quickly involves the joint. *Treatment.* Evacuate pus immediately, and splint to prevent deformity.

What is white swelling?

White swelling, or gelatinous arthritis, is a strumous inflammation of a joint, beginning usually as a (tubercular) synovitis, and characterized by slow course, with ultimate tendency to total disorganization of the part.

Swelling. Diffuse and somewhat elastic.

Pain. Gnawing in character, not very acute.

Color. Usually blanched.

Atrophy. Well marked.
Preternatural mobility. Readily detected.
Impairment, but not loss of function.

Give the treatment for white swelling.
1. Absolute rest, by means of fixed dressings kept on for months.
2. Tonics, stimulants, alteratives, cod-liver oil, quinine, iodide of iron.
3. Fresh air and good food in abundance.

Coxalgia.

What is coxalgia?
Coxalgia is a strumous arthritis of the hip-joint, occurring usually in persons under fifteen years of age. It is more common in boys than in girls, and is frequently tubercular.

Name the varieties of coxalgia.
1. *Femoral.* The disease begins in the upper epiphysis of the femur.
2. *Acetabular.* The floor of the acetabulum is first involved.
3. *Arthritic.* The disease begins as a synovitis.

Into what stages may coxalgia be divided?
1. *Inflammation. Flexion* and *fixation* of joint.
2. *Effusion. Flexion, abduction,* and *fixation,* with *apparent lengthening* from compensatory curvature of the spine.
3. Frequently *suppuration. Flexion, fixation, adduction,* and *inversion. Apparent* shortening, due to a compensatory curvature of the spine in the opposite direction. Backward luxation of femur may take place in this stage.

What are the early symptoms of hip-joint disease?
Pain, frequently referred to knee.
Tenderness, elicited by jarring the femur upward, or pressing suddenly inward upon the trochanter.
Limping, which may wear off in the evening.
Fixation, detected by attempting to flex, extend, and rotate

the femur, when the muscles resist and the pelvis is felt to move with the thigh. Place the patient on his back, upon a bed or table, and press the knee of the affected side downward till the popliteal space touches the supporting surface, the lumbar vertebræ can be felt arching upwards. Raise the thigh to a right angle with the pelvis, the vertebral arch disappears, and on further flexion, the pelvis on the affected side is raised from the table.

Flexion. The limb of the affected side is *slightly flexed* and *carried in advance* of its fellow, the latter bearing the weight of the body.

What symptoms denote the further extension of the disease?

Second stage. Pain is more intense, with "starts" at night (showing exposure of bone by erosion of cartilages). *Tenderness, limping,* and *fixation* are more marked. *Swelling* may be perceptible. *Atrophy* is apparent; *nates flattened; gluteo-femoral fold* less distinct than on the sound side, circumference of thigh and leg lessened. *Position.* Limb flexed, abducted, and everted, with pelvis lowered on affected side. *Failure* in general health.

Third stage. Position. Flexion, adduction, and inversion, the affected thigh crossing the other. Pelvis elevated on the diseased side. *Shortening, real* from wasting, and *apparent* from spinal curvature. *Suppuration* and *abscesses* common. *Hectic* with rapid emaciation.

How may you distinguish between the various forms of coxalgia?

The *arthritic* form approaches nearer to the type of an acute inflammation, with sharp pain in the hip-joint, swelling, etc. The *femoral* variety is characterized by starting pain most marked at the knee (obturator and anterior crural nerves), by shortening and luxation as the disease progresses, by abscesses pointing to outer part of thigh, below the trochanter.

Acetabular. Tendency to abscess most marked, may point from within the pelvis, over the nates, or above Poupart's ligament.

What is the prognosis in hip-joint disease?

Arthritic form is, in children, favorable. Femoral and acetabular forms more grave, especially the latter. In adults the prognosis is unfavorable.

What are the complications of hip-joint disease?

1. Suppuration. 2. Amyloid degeneration. 3. Tubercular meningitis.

How do you treat hip-joint disease?

In light and beginning cases, a fixation splint to the affected side (Agnew's, Thomas's, or a plaster bandage), a high-soled shoe (three inches) on the sound side, and a pair of crutches. For more serious cases, *rest in bed, with extension apparatus*, as in fractures, applied to the affected side, and counter-irritation, by means of blisters, over the inflamed joint. On disappearance of *all symptoms* get the patient up with high shoe, crutches, and splint, which must be continued for one year.

Constitutional treatment on general principles. Plenty of nourishing food and fresh air. Stimulants and tonics as required. *Cod-liver oil* and *syrup ferri iodidi*. Abscesses should be evacuated promptly by aspiration, or incision and drainage, under antiseptic precautions.

How do you treat anchylosis in a faulty position, following hip-joint disease?

By subcutaneous division of the neck of the femur by means of a strong narrow saw (Adams's), bringing the thigh into good position (extension), and treating as a fractured femur.

Continuous extension may succeed without an operation, in some cases.

Under what circumstances should the head of the femur be excised?

1. When it is necrosed and detached.
2. When other treatment has failed to check very free suppuration and rapid exhaustion of patient.
3. In some cases of displacement.

Under what circumstances is amputation justifiable in the treatment of hip-joint disease?

1. When there is extensive disease of the femur and free suppuration.
2. After excision which has not modified symptoms.

How do you distinguish between psoas abscess and coxalgia?

Psoas abscess can be felt as a fluctuating swelling, appearing *to the outer side* of the bloodvessels below Poupart's ligament, and traceable, through the abdominal wall, along the course of the psoas muscle. On marked flexion the pelvis does not move with the femur. Extension gives pain, referred to the loins.

Sacro-Iliac Disease.

Describe sacro-iliac disease.

Sacro-iliac disease is a strumous arthritis of the sacro-iliac joint, occurring in early life, and characterized by—

Pain over the affected joint, aggravated by coughing, straining at stool, or by lateral pressure.

Tenderness and swelling in the region affected.

Lameness appearing early.

Lengthening real, from downward displacement of os innominatum. *Suppuration*.

The *prognosis* is bad. *Treatment* as in case of hip-joint disease.

White Swelling of the Knee-Joint.

Describe white swelling of the knee-joint.

White swelling of the knee is usually a strumous (tubercular) affection, occurring in children, and characterized by—

Pain, slight at first, becomes starting.

Swelling, moderate at first, gradually increasing.

Tenderness, particularly marked on inner aspect.

Lameness, not producing entire disability for some time.

Displacement. Knee at first flexed, but as ligaments are softened and yield, there is a *backward* displacement and outward rotation of the tibia on the femur.

Crepitus, marked. *Undue mobility*, in a lateral direction.

Abscesses may form, opening externally, or the joint may become anchylosed.

Treatment. Fixation in good position, as for chronic synovitis and arthritis.

Rheumatoid Arthritis.

Describe rheumatoid arthritis (osteo-arthritis).

Seats. 1. Hip. 2. Shoulder. 3. Jaw.

Lesions. Absorption of cartilage, ulceration of bone surfaces with rarefaction, shortening of ligaments, and bony deposits in and around the joint. *Occurs* after middle life, usually in men.

Symptoms. Frequently bilateral; disability, some deformity, crackling, and atrophy.

Treatment. Local support, quinia, and general hygiene.

Loose Bodies in Joints.

What are the causes of loose bodies in a joint?

1. From altered blood-clot (fibrinous).
2. From hemorrhage into a synovial fringe, which subsequently organizes and is loosened.
3. From the gradual detachment of a synovial fringe.
4. In rheumatoid arthritis synovial fringes may be converted into cartilage, and become pediculated or loosened, or the nodular masses about the joint may project into the articular cavity.
5. As the result of injury, a portion of cartilage may be either chipped off or may, by a process of necrosis, be shed into the joint.

Knee-joint usually affected.

Give the symptoms of loose bodies in a joint.

Recurrence of attacks characterized by—

Sudden, agonizing pain, and *fixation of the joint* in slight flexion, followed by synovitis.

Detection of the body by manipulation; commonly found in the pouch over the external condyle of the femur.

How do you treat loose bodies in joints?

Radical. Secure the body in place by transfixing it with a strong needle; dissect it out, checking bleeding before opening the joint. If it has a pedicle, ligate. Close the wound, dress, and immobilize.

Palliative. Knee-cap.

Anchylosis.

What are the varieties of anchylosis or stiff joint?

True anchylosis is dependent on articular and intra-articular thickening and adhesions. *True anchylosis* may be *complete*, in which case the articular surfaces are united in part or throughout by bone. Rarely found except after traumatic arthritis.

Or it *may be incomplete*, motion being restricted by fibrous union between the joint surfaces, and thickening of the capsule.

False anchylosis is dependent on contractions and adhesions of the soft parts around the joints.

Give the treatment of anchylosis.

Incomplete or fibrous anchylosis. Passive motion and use. Application of splints, the angle of which can be changed. Continuous extension by means of weights. Forcible flexion and extension under anæstheties.

Complete or bony anchylosis. If the position is good, let alone, except in the case of the elbow, which should be excised. If the position is bad, osteotomy or resection.

DISEASES OF BONES.

Name the inflammatory diseases of the bones.
Periostitis, osteitis, osteomyelitis, epiphysitis.

Periostitis.

Describe periostitis.

1. *Simple local periostitis*, which may become *suppurative periostitis*, forming periosteal abscess.
2. *Diffuse infective periostitis.*

(1) *Local periostitis. Cause.* Local injury or extension of inflammation from other parts.

Pathology. Thickening of external fibrous layer, proliferation of inner osteogenetic layer, and inflammatory exudate loosening the periosteum from the bone. It may terminate in: 1. *Resolution.* 2. *Periosteal abscess.* 3. *Periosteal nodes* (particularly in chronic periostitis).

Symptoms. Pain. Intense, bursting, and worse at night.
Swelling of soft parts overlying.
Tenderness. Well marked on pressure. *Fever.*

In *suppuration*, symptoms are increased in severity; there are œdema, and discoloration of skin.

Treatment. Rest in bed, elevation, cold, opium for pain, leeches. Should pain and fever be unabated, or increase in twenty-four hours, free incision. If pus, open. For osteoplastic periostitis (periosteal nodes), oleate of mercury, subcutaneous section, or ablation by gouging.

(2) *Diffuse infective periostitis.*

Cause. Injury to a strumous subject.
Seat. Long bones; femur, tibia, humerus.
Pathology. Rapid septic suppuration, completely separating periosteum from bone.
Symptoms. High fever and *profound constitutional disturbance* rapidly running to a condition of septicæmia.

Deep-seated pain. Redness, puffiness, and *œdema* of the skin appear early.

Treatment. Early and free incisions. Antiseptic irrigation. Thorough drainage. Stimulants, tonics, and rich diet.

Osteitis.

Describe osteitis.

Cause. Injury, diathesis (scrofula, syphilis, rheumatism).

Pathology. Inflammatory exudation and cellular hyperplasia in the Haversian canals, with solution and removal of the bone substance. Haversian canals, lacunae, canaliculi become widened, and may disappear by coalescence. This constitutes *rarefying osteitis* or *osteoporosis*. The bones may yield to pressure and become greatly deformed, constituting *osteitis deformans*. If the inflammation is very acute, rapid proliferation causes strangulation of vessels and the bone dies in mass (*necrosis*), or by molecular death and discharge (*caries*). If inflammation is somewhat chronic, the absorbed bone is replaced by a new deposit, excessive in amount, and very dense (osteosclerosis or osteoplastic osteitis), or the inflammation may result in a localized collection of pus (*abscess of bone*).

Symptoms. As in periostitis. Osteocopic (starting) pains more marked. Tenderness on *tapping*. (Tenderness on pressure greatest in periostitis.) Limb heavier and more useless.

Treatment. As for periostitis. Hot fomentations of lead water and laudanum. Subcutaneous drilling. Trephine. Treat diathesis.

Osteomyelitis.

Describe osteomyelitis.

Definition. Inflammation of the marrow of the bone.

Cause. Traumatism. May occur primarily, or may be secondary to other affections of the bone.

Varieties. 1. Simple. 2. Suppurative. 3. Gangrenous.

1. *Simple osteomyelitis.* There is proliferation affecting the embryonic cells in the medulla and in the surrounding Haversian

canals and cancellous tissue, the fat disappears, the bone is absorbed. Granulation tissue is formed which may undergo *resolution*, may *organize* into *bone* filling the medullary canal (as in case of fractures), or may *suppurate*.

2. *Suppurative osteomyelitis.* May be *circumscribed* forming *bone abscess*, or *diffuse*, leading to extensive necrosis or pyæmia.

3. *Gangrenous osteomyelitis.* Due to a very high grade of inflammatory action, causing death by obstruction to circulation.

Complications of osteomyelitis. Caries, or bone ulceration. *Necrosis*, death of bone; this may be *central*, involving the inner laminæ only, *peripheral*, involving the outer laminæ, or *total*, involving the whole thickness of the shaft. *Separation of epiphysis. Inflammation of epiphysis. Pyarthrosis. Pyæmia.*

Osteomyelitis exhibits a tendency to spread *towards* the trunk.

Treatment. Simple osteomyelitis, as for osteitis.

Suppurative osteomyelitis. Open with trephine. If suppuration is extensive and associated with pyarthrosis (pus in joint), amputate.

Gangrenous osteomyelitis. Amputate.

Abscess of Bone.

Describe abscess of bone.

Nature. Usually strumous.

Cause. Due to rarefying osteitis, or the breaking-up of caseated tubercular masses.

Seat. Head of tibia usually (Brodie's abscess).

Symptoms. Boring persistent pain, worse at night. Tenderness especially marked on striking or tapping.

Treatment. Apply a rubber bandage and tourniquet and search for pus with a drill. Trephine; scrape, and chisel out all rough or carious bone. Pack with iodoform gauze, apply an antiseptic dressing and a splint.

Caries.

Describe caries.

Definition. Ulceration or molecular death of osseous tissue.

Pathology. As for rarefying osteitis. The surrounding bone is indurated, except in struma, when it is converted into a mass of fungous granulations.

Seats. Cancellated extremities of long bones. Often affects the joints secondarily.

Symptoms. Those of osteitis with abscess.

On probing, the softened, roughened, readily bleeding diseased area is detected. The discharge contains an excess of phosphate of lime.

Treatment. Remove the diseased bone by the curette, gouge, or osteotrite. When the detritus preserves its color in spite of washing, sound tissue is reached. Excision or amputation may be necessary.

Necrosis.

Describe necrosis.

Definition. Death of bone in mass.

Direct cause. Osteitis in any of its varieties.

Remote cause. Scrofula, syphilis, phosphorus, exposure to heat and cold, etc.

Necrosis may be *dry* (the ordinary variety), due to inflammatory strangulation, or *moist*, due to sudden death from injury.

Necrosed bone is dry, dirty yellow or brown, hard, and does not bleed when struck with a probe. When loosened it is thrown off as an *exfoliation*. The periosteum frequently retains its vitality, and throws out a sheath of new bone surrounding the dead portion, which, when it is entirely separated from the living bone and thus surrounded, forms a *sequestrum*, and is said to be *invaginated*. The sheath of bone investing the *sequestrum* is called the *involucrum*. The openings in the involucrum, through which the discharge makes its way to the surface, are called *cloacæ*. Dead bone is separated from the living by a process of granulation.

Sequestrum. Dead bone surrounded by living bone.
Involucrum. A shell of living bone surrounding a sequestrum.
Cloaca. Openings in an involucrum.
Symptoms. Those of bone inflammation, followed by free suppuration, with discharge of laudable pus; this continues for a long time, the abscess openings contracting down to sinuses.
Diagnosis. Made by feeling the hard, rough surface of dead bone with a probe.
Treatment. Nourishing food, tonics, fresh air, iodide of iron, and cod-liver oil. Sequestrotomy when the sequestrum is loose.

Tubercle.

Describe tubercle of bone.

Three forms. *Miliary tubercle, caseating tubercle,* and *scrofulous osteitis* (chronic rarefying osteitis). *May be local* (encysted) or *diffuse* (infiltrated); more commonly the latter.
Seat. Cancellated ends of long bones.
Common form. Scrofulous osteitis (tubercular nature cannot always be proven); occurs chiefly on hands, feet (strumous dactylitis), ends of long bones (abscess, or scrofulous arthritis), and *bodies of vertebra* (Pott's disease).
Symptoms. Those of osteitis, together with the signs of scrofulous diathesis.
Treatment. Air, good food, general hygiene, etc.
Counter-irritation, pressure, and splinting. When suppuration takes place, open, and remove entire disease area.

Syphilitic Bone Disease.

Describe the osseous lesions of syphilis.

Acquired. Gummata between periosteum and bone, forming periosteal nodes. These nodes chiefly affect the tibia, ulna, clavicle, and hard palate. Rarely, a diffused chronic form of inflammation causes syphilitic osteitis or sclerosis.
Congenital. In very young children *cranio tabes,* or wasting

of bone at the sites of decubitus, *i. e.*, behind the eminences of the parietal bones. *Alterations in the epiphyseal cartilage* making the bone *brittle* and *soft*. *Hutchinson's teeth*, and *Parrot's nodes* or *osteophytes*, appearing in the form of bony projections about the anterior fontanelle, and on the tibia and humerus.

Osteomalacia.

Describe mollities ossium or osteomalacia.

A disease characterized by general softening of the bones, rendering them liable to be bent or broken.

Occurs during and after adult life, mostly in females.

Pathology. Rarefaction and absorption of bone, advancing from the centre outward. Replacement of medullary tissue by a dark-red, semi-fluid material.

Symptoms. Obscure pain in the bones and malaise. Phosphates in the urine. Fractures, deformity.

What is fragilitas ossium?

A brittleness of bone dependent on fatty degeneration.

Pott's Disease.

What is Pott's disease?

Pott's disease is an *angular* deformity of the spine caused by caries of the vertebræ or the intervertebral cartilages.

Give the pathology of Pott's disease.

Usually due to a tubercular osteitis which affects the bodies of several vertebræ simultaneously; these becoming softened, yield to the superimposed weight, thus causing deformity. There may be no pus formation, the inflamed area being removed by *interstitial absorption*, the pus may become encysted and caseated, or, more commonly, may appear as a *cold abscess*. The cord is rarely injured, the deformity being so gradual that it accommodates itself to its new course.

Anchylosis, which is a reparative effort, goes hand in hand with the disease, new bony arches being thrown out between the vertebrae. Pott's disease occurs most frequently in childhood, and is commonly found in the dorsal and cervical regions.

Give the symptoms of Pott's disease.
1. *General failure* in health.
2. *Rigidity of spine.* Detected by getting the patient to pick an object from the floor, to rise from a dorsal recumbent posture, or to turn from the back to the belly. In consequence of rigidity and tenderness, the *gait is tottering, shuffling,* and *uncertain.*
3. *Pain* and *tenderness,* elicited at times by jarring the head or by inducing the patient to jump from a chair or step. May be found by direct pressure. There is a constant tendency to support the back; the patient will frequently lie down, or, if sitting, will support the weight of the shoulders on the thighs.
4. *Reflex irritation. Lumbar disease* is frequently attended with colicky pain, irritation of the bladder, and incontinence of urine. *Dorsal disease* is characterized at times by a *grunting respiration. Cervical disease* may cause torticollis, choreic movements of the neck muscles, or difficulty in deglutition.
5. *Deformity.* Undue prominence of spinous process causing a backward projection.
6. *Abscesses.*
7. *Paresis or paralysis.*

In what directions do the abscesses of Pott's disease point?
Cervical region. Post-pharyngeal abscess may be formed, or the pus may pass outward between the longus colli and scaleni muscles, appearing behind the sterno-cleido-mastoid, or it may pass downward.

Dorsal region. Pus may pass directly backward, or form psoas, iliac, or lumbar abscess.

Lumbar region. Lumbar abscess, appearing to outer side of quadratus lumborum. *Psoas* or *iliac* abscess.

Give the treatment of Pott's disease.
Constitutional, as for strumous affections.

Local. Rest. In the early stages rest in bed. *Plaster jacket* with either entire or partial confinement to bed.

Abscesses must be opened as soon as detected. Open psoas abscesses

How is the plaster jacket applied?

Bandages two and one-half or three inches wide, seven yards long, made of gauze, mull, or crinoline. Rub dry plaster of Paris thoroughly in the meshes of each bandage as it is rolled. Place on the patient a clean thin summer undershirt, pad all bony projections with cotton, put over the abdomen next to the skin a "dinner pad" (a folded towel), suspend the patient by the head and shoulders, wet the bandages, and apply them so that the expanded basin of the pelvis is caught below and the support comes well up beneath the axilla of each side. Remove the dinner pad when the bandage hardens.

Rickets.

Define rickets.

Rickets is a constitutional disease of childhood, characterized by lesions of the osseous system, and a tendency to amyloid degeneration of the viscera.

Etiology, defective or unsuitable food.

Give the pathology of rickets.

Increased cell-growth, with deficiency of earthy matter. Enlargement of epiphyseal cartilages. Thickening of periosteum. Softening and distortion of the shafts of the bones.

Give the symptoms of rickets.

Premonitory. Delayed dentition, restlessness at night, sweating about the head, abundant urine loaded with phosphates.

Of the developed disease. Deformities. Such as—

1. *Pigeon-breast*, with beaded ribs from enlargement of costo-chondral junction.

2. *Lateral* or *antero-posterior* curvatures of the spine.

3. *Bent legs* or *arms* with *rounded enlargements* at the ends of the long bones.

As a frequent complication we have *bronchitis*, serious on account of the yielding nature of the chest walls.

Treatment. General hygiene, nourishing diet, cod-liver oil, lactophosphate of lime, iron, syrup. hypophos. comp.

Hæmophilia.

Describe hæmophilia.
Hæmophilia is a congenital and habitual hemorrhagic diathesis, in virtue of which persistent bleeding may occur, of itself, or from the slightest wound.

Treatment. Compresses saturated in Monsel's solution, strong pressure, ergot, acetate of lead, and other hæmostatics.

Struma.

What is struma?
Struma or scrofula is a defective bodily condition characterized by a tendency to the development of *chronic (tubercular)* inflammations of the bones, joints, and lymphatic glands.

What are the characteristics of scrofulous inflammations?
1. They develop at an early period in life.
2. They are chronic in type.
3. They occur chiefly in phthisical families.
4. They exhibit a marked tendency to pass on to suppuration and caseation.
5. They are prone to appear in certain regions. Example, cervical adenitis.

Give the treatment of scrofulous inflammation.
Constitutional. Generous diet, fresh air and sunshine, cod-liver oil, iodide of iron.

Local. Active counter-irritation, pressure, operative procedures.

Curvature of the Spine.

Describe spinal curvature.
The curvature may have its convexity directed forward, backward, or to the side.

The cause of curvature is long-continued, unequal compression of the intervertebral cartilages.

Forward curvature, or *lordosis*, is usually found in the lumbar region, and is simply an exaggeration of the normal curve, compensatory to some deformity or diseased condition, such as ricket, congenital femoral luxation, coxalgia, etc.

Backward curvature, or *kyphosis*, usually appears as an exaggeration of the normal dorsal curve. It is the result of debility, rickets, or occupation requiring constant stooping.

Treatment. In the young, friction, massage, deep breathing, exercises for back muscles, braces which are comfortable only when the shoulders are held back.

Lateral curvature, or *scoliosis*, develops most frequently in girls, between the ages of 14 and 18. There are usually two curves with their convexities turned in opposite directions. The vertebræ are rotated on their vertical axes, their spinous processes pointing towards the concavity of the curves.

Causes. Inequality in the length or strength of the legs; one-sided position or use of the body; contractions following empyema or paralysis of spinal muscles of one side. These causes are rendered more operative by debility, or a strumous or rachitic diathesis.

Symptoms. Sense of fatigue and pain in back and shoulder when sitting, or on first lying down. Wing-like projection of scapula (dorsal curvature is usually toward right), and undue prominence of the iliac crest of the affected side, with projection of the breast on the opposite side. Curvature may be detected by marking the spinous processes, though it must be remembered that the amount of deformity is much greater than is indicated by this test.

Treatment. Change in habits or occupations which can act as exciting causes. Massage, friction, and electricity to the muscles of the back, systematic gymnastic exercises, suspension followed by rest in the recumbent position. If deformity increases, apply a plaster-jacket.

HERNIA.

What is a hernia?

The protrusion of a viscus through an abnormal opening in the walls of the cavity in which it is contained.

As applied, hernia is synonymous with *rupture*, and indicates protrusion of the *abdominal viscera* through abnormal openings in the *parietes*.

What are the essential parts of a hernia?

1. The sac. 2. The contents.

Describe the sac.

The sac may be (1) *congenital*. Found only in umbilical and inguinal regions; consisting of a pouch of peritoneum ready to receive the hernia. (2) *Acquired*. Developed by gradual stretching of the parietal peritoneum. This is the form of sac ordinarily found.

The *formation of the sac*. Pressure of abdominal contents upon the parietal peritoneum may cause a bulging of the membrane where it is poorly supported, as at the internal inguinal ring; the peritoneum yields, and the bulging is developed into a pouch which fills the inguinal canal; escaping from the external ring its base is less supported, and it forms a pyriform swelling, consisting of—(1) *The neck*, at the internal ring. (2) *The body*, the main part of the sac. (3) *The fundus*, or wide extremity. As the peritoneum is dragged downward it becomes puckered at the neck.

During the stage of (1) Formation, this puckered neck exerts no constriction upon the hernial contents.

Stage 2. *Organization*. These puckerings become adherent, and the surrounding subserous fat is indurated.

Stage 3. *Contraction*. The neck of the sac contracts and may become obliterated, or may cause strangulation if the gut be protruding.

The sac, at first smooth, becomes thickened, contracts, adheres, and is irreducible; at times it sends off diverticula or secondary sacs.

How are hernias classified in regard to the contents of the sac?

1. *Epiplocele.* Containing omentum only, most common on left side.
2. *Enterocele.* Containing intestine only, usually ileum.
3. *Entero-epiplocele.* Containing both omentum and gut.

Further we may have *cystocele* (bladder), *cœcocele* (cæcum), *gastrocele*, etc.

What are the causes of hernia?

1. *Predisposing.* *Sex*, males. *Heredity.* *Age*, young. *Lengthened mesentery.* *Structural defects* (congenital). *Occupation.* *Abnormal conditions*, such as a protracted cough, operations on the abdomen, and muscular relaxation.
2. *Exciting.* Muscular contraction.

What are the common seats of hernia?

In the inguinal, femoral, and umbilical regions.

What are the varieties of hernia in regard to their condition?

(*Clinical varieties.*)

1. *Reducible.* Most common form, the contents can readily be returned into the abdomen.
2. *Irreducible.* Contents cannot be reduced into abdomen.
3. *Obstructed or incarcerated.* The contained bowel becomes obstructed by its contents.
4. *Inflamed.* There is inflammation or localized peritonitis of sac and contents.
5. *Strangulated.* Subject to a constriction not only obstructing the bowel, but seriously interfering with its circulation.

Reducible Hernia.

What are the symptoms of reducible hernia?

1. *Enterocele.* A smooth, regular, round tumor in a hernial region, often to be traced through the hernial canal, *larger* on *standing* than on *lying down*. *Tympanitic* on percussion, *gurgles* when manipulated. *Disappears with a flop* when pressed inwards. Presents *succussion* (an expansile push) on coughing. Local weakness, dragging pains, and irregular dyspepsia.

2. *Epiplocele.* No tympanites, no flop, no gurgle; the symptoms the same but less marked. Doughy and uneven on palpation.

Give the treatment for reducible hernia.

1. *Palliative.* 2. *Radical.*

Palliative. Truss, consisting of pad and spring. Pad must be *slightly* convex, and large enough to cover the external opening and the canal through which the hernia descends. The spring must so act on the pad that the pressure is just sufficient to keep the hernia up.

To test a truss, let the patient stoop, cross the legs, and cough, sitting on the edge of a chair with the body leaning forward and legs widely separated.

To measure for a truss. (Inguinal or femoral.) From lower border of hernial opening to the anterior superior spine of ilium of same side, from this point around the body one inch below crest of ilium to other iliac spine, thence to upper part of hernial opening.

Directions for use. Immediately remove truss if hernia should come down. Bathe the skin beneath the pad with whiskey and alum on taking off the truss, and before replacing it. Take off *after* lying down and replace *before* rising.

Radical cures. The various operations devised for this purpose have in view: 1. *Obliteration of the neck of the sac* either by ligature, or stitches, or by plugging it with the invaginated fundus. 2. The *obliteration of the canal;* and 3. *The closure of the external and internal rings.*

Irreducible Hernia.

What are the causes of irreducible hernia?

Temporarily irreducible, from slight distension with fæces or gas.

Permanently irreducible, from the bulk of the tumor, constriction of the neck of the sac, adhesions within the sac, fatty enlargement of prolapsed omentum.

How do you treat irreducible hernia?

Temporarily irreducible, as for incarcerated.

Permanently irreducible. If very large, apply a bag truss, if moderate in size, fit a truss with a concave pad; advising, in all cases where there is pain or discomfort, an operation for the *radical cure* of the hernia.

Incarcerated Hernia.

What are the symptoms of obstructed or incarcerated hernia?

Occurs mostly in *irreducible hernia*, particularly in such as contain colon. Constipation is a strong predisposing factor.

1. Tumor is *enlarged and slightly tender*. Liquid and gaseous contents may be pressed out, and doughy fæces detected.
2. There is some pain, with distension of the stomach, constipation, nausea, and vomiting.
3. The constitutional symptoms are of moderate severity.
4. There is impulse on coughing.

How do you treat incarcerated hernia?

Treatment. Rest in bed, cracked ice by the mouth, complete relaxation by position. Apply an ice-bag to the hernia, and give opium if there is pain. Open the bowels by purgative enemata, followed by castor oil as soon as the tumor is diminished in size. If symptoms of obstruction persist, perform herniotomy.

Inflamed Hernia.

Describe inflamed hernia.

Cause. Injury to a small irreducible hernia, usually inflicted by a badly fitting truss.

Symptoms. Chiefly those of acute local inflammation. Redness, heat, pain, swelling (nodulated if epiplocele, sac contains fluid if enterocele), impulse on coughing. Fever, vomiting, and constipation of moderate severity. Wind passed by bowels.

Treatment. Opium if great pain. Rest in bed with local

relaxation by position. Ice-bag to the inflamed part. Opening enema (soap and water Ojss). Gentle purgation when inflammation subsides.

Strangulated Hernia.

What are the causes of strangulated hernia?

1. Sudden descent into the sac of an irreducible hernia of an additional mass of omentum or intestine.
2. Sudden descent of a hernia long retained by a truss.
3. Parietal constriction about the opening of a hernia suddenly produced by violent effort.

Where is the seat of constriction?

1. *At the neck of the sac.* At times in the body of the sac, from hour-glass constriction.
2. *Entirely within the sac.* Due to bands of lymph, or a rent in the omentum.
3. *Entirely without the sac.* In small hernia suddenly produced by violent effort.

What changes take place in strangulated hernia?

Bowel is grooved by constriction, becomes œdematous, ecchymosed, red deepening into purple, *loses its lustre, becomes harsh, sticky, non-elastic,* and *dirty black.*

Sign of local death—*loss of lustre and elasticity.*

May rupture into the sac, or at the line of constriction. Inflammatory adhesions mostly prevent fæcal extravasation into the peritoneal cavity.

Sac, attacked by inflammation, effuses serum.

What are the symptoms of strangulated hernia?

1. *Tumor becomes more tense,* somewhat *duller* on percussion, *tender* at the *neck* of the sac, and *gives no succussion on coughing.*
2. Abdominal pain, with sense of constriction about umbilicus.
3. *Vomiting,* frequent and persistent; first, contents of stomach, then bile, finally fæces.
4. *Obstinate constipation.*

5. *Rapid loss of strength; small, rapid, compressible pulse; dry, brown tongue.* Very *little urine* passed, it may contain albumen and indican, and be deficient in chlorides.

Gangrene is denoted by *cessation of pain and vomiting*, and rapid development of symptoms of *collapse*.

What is Littré's hernia?

A hernia involving only a portion of the circumference of the bowel. Though the pouch is strangulated, there is not absolute internal obstruction.

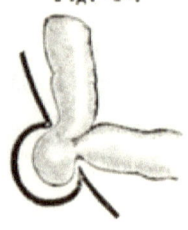

Fig. 40.

Littré's hernia.

What are the symptoms of Littré's hernia?

As for strangulated hernia, but less marked; vomiting *not* stercoraceous, constipation *not* absolute. Tumor is small, and gangrene rapidly develops; hence the treatment is *early herniotomy*.

What are the principal points in the diagnosis of strangulated hernia?

1. Stercoraceous and persistent vomiting.
2. Absolute constipation.
3. Great constitutional depression.
4. Absence of *succussion*, or impulse on coughing.

How do you treat strangulated hernia?

Rest. Relaxation of parts by position. *Taxis. Herniotomy.*

How do you employ taxis?

Anesthetize, and fully relax by position (flexion and adduction of thigh for femoral or inguinal hernia). The head and shoulders should be low, the pelvis elevated. Define the neck of the sac with the thumb and forefinger of the left hand, then with the fingers of the right hand draw the sac down a little, and by a kneading, rolling, compressing movement press the gut in a direction corresponding to its line of the descent.

In *oblique inguinal* hernia the pressure must be outwards, upwards, backwards.

In *femoral* hernia first slightly downwards till falciform process is cleared, then directly backwards towards pubic spine.

Taxis failing in five to eight minutes, perform herniotomy.

Under what circumstances must taxis be avoided?

1. Very acute cases, as in hernia of sudden development, from violent muscular action.
2. Where symptoms of strangulation have existed for several days.
3. Where the strangulated gut was previously irreducible.
4. Where the gut is gangrenous.

What accidents may occur in the employment of taxis?

1. Reduction *en masse* or *en bloc*. The hernia, *together with its sac*, is pushed directly inward, the strangulation being in no way relieved. Denoted by slow, difficult, forcible reduction not accompanied by gurgle or flop, and by *persistence of symptoms*.

2. Reduction *en bissac*. The bowel is pressed into a congenital diverticulum or pouch, running from the body of the sac below or beneath the abdominal muscles. Symptoms the same as reduction en bloc.

3. Reduction through a rupture in the neck of the sac, the hernia escaping into the subserous cellular tissue.

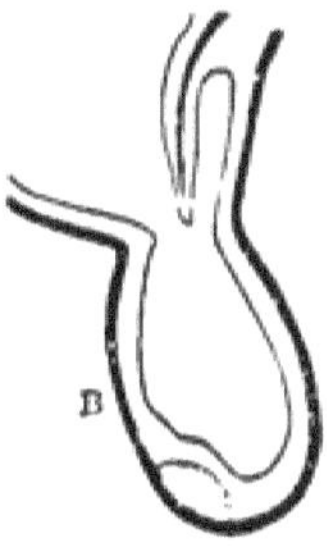

Fig. 41.   Fig. 42.

Reduction en bloc.   Reduction en bissac.

These three forms are usually classed as *reduction en bloc*. *Treatment.* Cut down, secure the sac, open it, and divide the constriction at the neck.

4. *Rupture of intestine.* Rapid collapse, no *gurgle*.

Under what circumstances may symptoms persist after complete reduction?

1. Paralysis of bowel.
2. Internal strangulation (causes within sac).
3. Acute peritonitis.

What treatment should follow reduction by taxis?

Compress and bandage locally. Absolute rest, milk diet; opium to quiet pain. If no inflammatory symptoms, open bowels by castor oil or purgative enemata the fifth day.

What treatment should follow continuance of symptoms after reduction?

Exploratory laparotomy, and careful search for causes of obstruction.

Describe herniotomy.

Empty bladder and rectum. The antiseptic method must be carried out to its minutest details. Shave the seat of operation, pinch up a fold of skin and transfix, cutting outward and making an incision about three inches long. Divide the successive layers of tissue on a grooved director till the sac is reached. The sac is *tense, rounded, bluish, with arborescent vessels*. Pinch up a small portion with forceps, and notch; a straw-colored or blood-stained serum escapes. Open freely with scissors, pass the finger up to the seat of constriction, slip the nail under the resisting band, pass a probe-pointed hernia knife along the finger, turn the edge forward, and divide the stricture. If the gut is in good condition, return; then restore the mesentery, and sew across the neck of the sac, removing its body, or do a formal radical operation. Insert a drainage-tube, close the external wound, and apply antiseptic dressing, compress, and bandage.

No food for twenty-four hours, then milk diet. Enema in two days.

How should the intestine be managed?

Return if it be *smooth, glistening, and elastic*, even though there be great discoloration and ecchymosis. *Draw down a little more of the gut and inspect the line of constriction* before returning. This is a common seat of perforation.

All manipulations must be practised with great gentleness.

A dull black, sodden, sticky bowel is beyond hope of recovery and must not be returned.

How do you treat gangrenous bowel?

Lay open the sac, carefully relieve the stricture, incise the gut if it is greatly distended, dress with charcoal poultice.

This leaves either a fæcal fistula (a small aperture discharging fæces), or an artificial anus (a double-barrelled opening).

If only a limited portion of the bowel is gangrenous, excise, and unite the healthy tissue with *Czerny's suture*, the first row including only the edge of the serous membrane, the second (Lembert's) starting one-half inch from the edge of the wound, and including a quarter of an inch of all the coats of the bowels except the mucous membrane.

How do you treat a fæcal fistula or an artificial anus?

The *fæcal fistula* frequently closes spontaneously; if not, a plastic operation may be performed, or it may be treated as an artificial anus.

In *artificial anus* the spur or partition formed by the anterior projection of the posterior wall of the bowel may be ulcerated through by means of Dupuytren's enterotome, after which the external opening may be closed by a plastic operation; or the intestine may be detached from the abdominal wall, drawn out, freshened, and united by Czerny's suture. Prepare by twenty-four hours' light diet, and thorough washing out of the bowels.

How should the omentum be managed?

If acutely strangulated, clamp, excise, secure the bleeding points, and return the stump to the abdominal cavity. If adherent, excise. Omentum must not be left in the sac.

How do you treat adhesions?

Break down recent adhesions. Apply two ligatures, and cut between old vascular adhesions.

How do you treat the sac?

Dissect it out, suture across the neck, and excise below the suture line.

What is the after treatment?

No food for thirty-six hours. Morphia hypodermically for pain. Stimulants, if necessary, by the rectum. Open bowels by an enema the seventh day. Remove the drainage-tube in forty-eight hours, the sutures on the fourth day. Keep up firm pressure by means of bandages. In one month apply a truss and get the patient out of bed.

Special Hernias.

What are the varieties of hernia in regard to position?

Diaphragmatic, Inguinal, Femoral,
Epigastric, Obturator, Lumbar, Perineal,
Ventral, Umbilical, Ischiatic, Pudendal.

Inguinal Hernia.

What is the most common variety of hernia?
Inguinal hernia.

Name the varieties of inguinal hernia?
1. Acquired.

Complete. When the hernia has passed through the external ring.

Incomplete. When the hernia is still in the inguinal canal, called also *Bubonocele*.

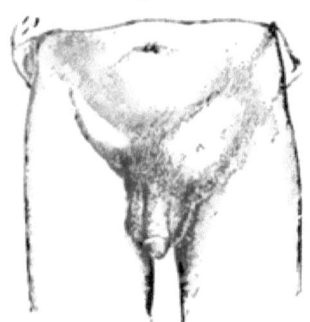

Fig. 43.

Inguinal hernia.

Oblique. Commonest variety. The hernia passes to the *outer side* of the epigastric artery, and if complete, through *the two rings and the canal*.

Direct. The hernia passes to the *inner side* of the epigastric artery and through the *external abdominal ring* only.

Further, a complete inguinal hernia reaching the scrotum is called *scrotal*, or the labium, in woman, is termed *labial*.

Rarer forms, depending upon congenital defects, are—

1. *Congenital hernia.* In this the peritoneal process (vaginal process), accompanying the testis in its descent, remains an open pouch and receives the gut.

2. *Hernia into the funicular portion of the vaginal process (infantile hernia).* This implies the same condition as before, except that the proper tunic of the testis has become closed, the funicular (cord) portion of the process alone remaining patulous.

3. *Encysted hernia.* The ventricular orifice of the peritoneal pouch is closed, the funicular and testicular parts remaining open. This hernia is of gradual formation. It invaginates the existing pouch and carries an additional layer of peritoneum with it, making three layers of serous membrane to be cut through.

Fig. 44.
Congenital hernia.

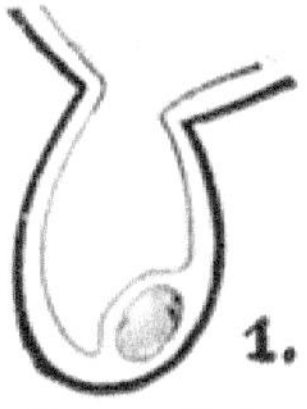

Fig. 45.
Infantile hernia.

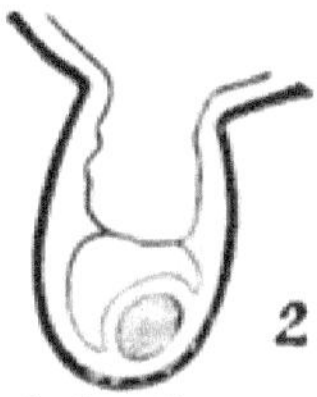

Fig. 46.

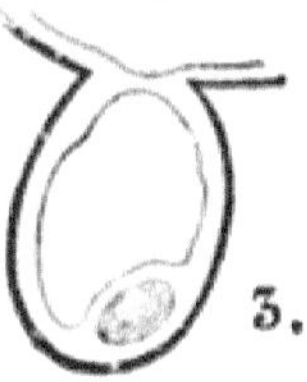

Fig. 47.

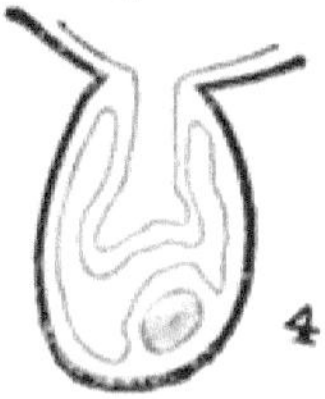

Encysted hernia.

Describe the inguinal canal.

The inguinal canal is an oblique passage through the anterior abdominal wall, lying parallel to Poupart's ligament and above it. It begins at the internal ring, ends at the external ring, and is one and one-half inches long. It transmits the spermatic cord in man, the rounded ligament in woman. It is bounded—

In front, by the external oblique, internal oblique (outer third), cremaster muscles.

Behind, by the conjoined tendon (inner third), transversalis fascia, triangular ligament, sub-peritoneal tissue, deep epigastric artery, and peritoneum.

Above, by the arch made by the internal oblique and transversalis.

Below, by Poupart's ligament and the transversalis fascia.

Describe the internal abdominal ring.

The *internal* abdominal ring is an oval opening situated in the transversalis fascia, one-half inch above the middle of Poupart's ligament. Above and external to it lie the arched fibres of the transversalis, below internally the deep epigastric artery. From its circumference a thin funnel-shaped membrane, the infundibuliform fascia, is continued around the cord.

Describe the external abdominal ring.

The external abdominal ring is a triangular aperture in the fascia of the external oblique muscle, bounded below by the crest of the pubis, above by the intercolumnar fibres. Internally and above by the internal column inserted upon the front of the pubic symphysis. Externally and below by the external column, inserted upon the pubic spine.

Describe Poupart's ligament.

Poupart's ligament is that portion of the fascia of the external oblique muscle extending from the anterior superior spinous process of the ilium to the pubic spine. In the lower portion it forms the external column of the external ring; a backward reflection from the pubic spine to the pectineal line forms *Gimbernaut's ligament*. A band of tendinous fibres continued from its attachment to the pectineal line up and in towards the linea alba forms the *triangular ligament*.

What is the cremasteric fascia?

It consists of the muscular fibres carried down from the internal oblique by the testicle in its descent; they form a series of loops covering the cord.

What are the coverings of an oblique inguinal hernia?

Skin, two layers of superficial fascia, intercolumnar fascia (from columns of external ring), cremasteric fascia (from canal), infundibuliform fascia (from internal ring), peritoneum (true sac).

Name the coverings of a direct inguinal hernia.

Skin, superficial fascia, intercolumnar fascia, conjoined tendon, transversalis fascia, and peritoneum.

If the hernia passes to the outer side of the conjoined tendon, this structure is replaced as a covering by the cremasteric fascia.

What effect has a long-standing inguinal hernia upon the length of the canal?

The internal ring is dragged down till it lies almost directly behind the external ring.

What is the relation of the cord to inguinal hernia?

Below and behind.

In what direction should the incision be made in relieving the stricture of an inguinal hernia?

Upward and outward, parallel to Poupart's ligament.

Describe congenital hernia.

The testis, in its descent into the scrotum, is accompanied by a peritoneal pouch. The pouch becomes occluded at two points, the internal ring, and the top of the epididymis. The portion between these two points occupies the whole of the inguinal canal; it shortly shrinks, and is transformed to a fibrous cord.

If the peritoneal process remains patent throughout, we have the condition which gives rise to *congenital hernia*.

If it is occluded at the lower end, *hernia of the funicular process* (*infantile hernia*).

If it is occluded at the upper end only, and the occluding septum yields, we have *infantile hernia*.

How do you diagnose these forms of hernia?

Congenital and funicular hernia (infantile) usually occur in early life, are of sudden development, become *complete* at once, *do not drag down the internal ring*. They are very prone to become *strangulated*, and are *difficult to reduce*.

The congenital hernia *intimately surrounds* the testicle; all other forms of hernia lie above it.

The encysted hernia cannot be diagnosed before cutting; then it will be found to *have a double sac*.

Congenital hernia may be associated with undescended testicle. In this case it will protrude outward along the fold of the groin.

Prognosis of congenital hernia is good.

With what affections may inguinal hernia be confounded?

Varicocele, hydrocele of the cord, congenital hydrocele, and enlarged inguinal glands.

How do you diagnose hernia from varicocele?

Varicocele feels *soft, doughy,* and like a *bunch of worms* to the fingers. Disappears on lying down, to appear again on standing, but first *enlarges at the bottom* of the scrotum. If it is made to disappear, and the finger is placed over the external ring, it will appear *more quickly* than before. No gurgling, no tympanites, slight succussion. An omental hernia may feel doughy, but not like a bunch of earth-worms, the enlargement comes from above, and if reduced, the finger placed over the external ring will prevent it from reappearing.

How do you diagnose inguinal hernia from other affections of the same region?

Hydrocele of the cord is translucent, enlarges like varicocele from the bottom, and fluctuates. It has neither gurgling nor tympanites.

Undescended testicle. Absence of gland on affected side, *hard* tumor in inguinal canal, *sickening pain* on pressure.

Enlarged inguinal glands. Direction of tumor *oblique to long axis of canal.* It is hard, very painful, and the skin is reddened. *Tumor freely movable at first.* Hernia lies in the long axis of the inguinal canal, is soft, is not painful, the skin is normal, the tumor lies very deep, and is immovable.

Femoral Hernia.

Describe the femoral canal.

The femoral or crural canal is a narrow interval below Poupart's ligament, between the femoral vein and the crural sheath (sheath of the vessels).

It is one-quarter to one-half inch long, extending from the *femoral ring* to the upper border of the *saphenous opening.* The *septum crurale* closes the canal at the femoral ring, the *cribriform fascia* at the saphenous opening.

Describe the femoral ring.

The femoral ring lies between Poupart's ligament above, the pubis and pectineus muscle below, with Gimbernat's ligament to the inner side, the femoral vein to the outer side. It is oval in shape, about one-half inch in diameter, and is closed by the *septum crurale* and a lymphatic gland.

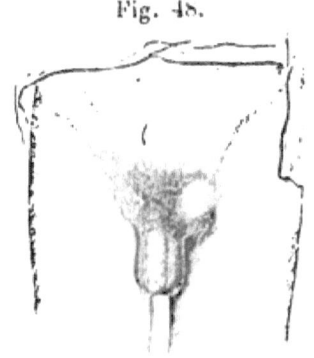

Fig. 48.

Femoral hernia.

Describe the saphenous opening.

The saphenous opening, formed by a reflection of the fascia lata beneath the femoral vein, is an oval-shaped aperture, one and one-half inches in length, one inch in breadth, situated beneath the inner portion of Poupart's ligament.

Its upper and outer margin, sharply defined and semilunar in shape, passes *in front* of the vessels and is inserted into the pubic spine and pectineal line. It is called the *superior cornu of the falciform process*. Its lower and inner margin forms the *inferior cornu of the falciform process*.

The inner margin is formed by the fascia passing to the pectineal line, curving upwards and *behind* the femoral vein, covering in the pectineus muscle. This portion of the ring is not sharply defined.

What are the boundaries of the femoral canal?

Anterior. Poupart's ligament, transversalis fascia, falciform process of fascia lata.

Posterior. Iliac fascia, pubic portion of fascia lata.

Internal. The junction of the transversalis and iliac fascia, forming the inner wall of the crural sheath, Gimbernat's ligament.

External. The septum covering the femoral vein.

What are the coverings of femoral hernia?

Skin, superficial fascia, cribriform fascia, crural sheath, septum crurale, peritoneum.

Where is the gut commonly strangulated in a femoral hernia?

Gimbernat's ligament. Superior cornu of falciform process, or Hay's ligament. (Agnew.)

What important structures lie near the femoral ring?

1. Spermatic cord, just above the superior margin.
2. Epigastric artery, passes above to the outer side.
3. Obturator artery, may curve across the upper and inner border.
4. Femoral vein to the outer side.

How do you distinguish femoral from inguinal hernia?

Femoral hernia, traced upward towards its neck, is found to pass to the outer side of the pubic spine. Inguinal hernia passes to the inner side.

In what direction should you cut in relieving the constriction of a strangulated femoral hernia?

Upward and inward, using a blunt-pointed knife with a dull edge.

How do you distinguish femoral hernia from a psoas abscess?

They both give succussion, and disappear on pressure or recumbency. Psoas abscess comes down to the *outer* side of the vessels, gives the signs of the diseased condition by which it is caused, and fluctuates. It can be *traced above Poupart's ligament*. Hernia appears to the *inner* side of the femoral vessels and has the characteristic signs. It cannot be traced above Poupart's ligament.

Umbilical Hernia.

What are the varieties of umbilical hernia?

1. *Congenital*, depends upon imperfect closure of the ventral plates, the sac extends into the cord and has been tied by the accoucheur.
2. *Acquired*, depends upon yielding of the abdominal cicatrix. This is the commonest variety of umbilical hernia, both in infants and adults.

What are the coverings of an umbilical hernia?
 Skin, superficial fascia, linea alba, sac.

How do you treat umbilical hernia?
 In infants, draw the recti muscles together, strap tightly, and apply a binder or bandage. In adults apply a protecting concave truss.

Where should the incision for relief of strangulated umbilical hernia be made?
 In the linea alba, beginning a couple of inches above the upper margin of the hernia. The parietal tissues are often very thin.

INTESTINAL OBSTRUCTION.

Give the causes of acute intestinal obstruction.
1. Congenital malformation, imperforate anus, etc.
2. Impaction of foreign bodies and gall-stones.
3. Invagination or intussusception.
4. Volvulus or twisting, commonly dependent on mesenteric elongated.
5. Internal strangulation, or constriction of the bowel by bands or diverticula having no structural connection with the circumference of the constricted gut.

Symptoms of acute intestinal obstruction may also appear in *enteritis*, peritonitis, and perityphlitis; or in chronic obstruction.

Give the symptoms of acute intestinal obstruction.
Pain, often intense and localized. *Vomiting*, gastric, bilious, intestinal, and finally fæcal. *Constipation*, absolute. Abdomen *swollen, tender, tympanitic*. *Peristalsis* increased, causing borborygmus and gurgling. *Great vital depression*. *Small, rapid pulse*. *Temperature* may be normal or subnormal till just before death, which commonly occurs in from seven to ten days.

How may the seat of acute intestinal obstruction be inferred?
The probability of the small intestine being involved is in direct proportion to the acuteness of the pain and the rapidity of the course. Early and severe vomiting, scanty urine, and early distension all point to small intestine.

What are the causes of chronic obstruction?
Fæcal accumulation, stricture of the bowel, glueing of the intestines together from chronic peritonitis or cancer, abdominal tumors.

Give the symptoms of chronic obstruction.
Constipation; *abnormal* distension very slowly developed; *vomiting* comes on slowly or may be absent; *pain* rarely acute; *constitutional depression* not marked.

What are the special characteristics of intussusception?

This is the common form of acute obstruction in infancy and childhood. Usual seat, ilio-colic valve. It is characterized by *tenesmus* and passage of *mucus and blood*.

Sausage-shaped tumor usually to the *left* side of the abdomen.

On examination per rectum the invaginated gut may be found.

Give the treatment for intussusception.

Inflation *per* rectum with air or water; inversion; gentle kneading of the bowels.

Laparotomy, and reduction by kneading and drawing down the sheath or outer tube. If reduction is not possible, make an artificial anus, or cut off the intussuscepted part, and suture together the two ends of the bowel.

What are the special characteristics of internal strangulation?

Occurs during adolescence or early adult life.

Patient has been previously healthy, symptoms following a *blow* or a *straining* effort.

Symptoms *very acute*. *Severe pain* referred to umbilicus with *intense prostration* or syncope. There is no peristalsis, no tumor.

What are the special characteristics of volvulus?

Occurs in advanced life.

Seats. Sigmoid flexure of colon, and in the neighborhood of the ilio-cæcal valve.

Symptoms are characterized by *extreme rapidity* and *severity*.

Give the treatment of acute intestinal obstruction.

Make most careful search in all hernial regions for strangulation. Keep the patient in the recumbent position. Give liquid nourishment and in minimum quantity. Morphia gr. ¼ every three to six hours, as required to relieve pain. Hot fomentations to the belly. Cocaine, hydrocyanic acid, etc., for vomiting. If, after a reasonable time (one to three days, according to the severity of the symptoms), there is no change for the better, *laparotomy*, with further measures adapted to the relief of the obstruction.

Give the treatment of chronic intestinal obstruction.
Enemata. If from impaction of fæces, break up mechanically and remove. If from malignant trouble, or stricture, excision, with circular enterorraphy or artificial anus.

What is laparotomy?
Opening the abdominal cavity.
Incision. Linea alba, midway between pubes and umbilicus, large enough to admit the fingers. Stop all bleeding before opening peritoneum. *Explore* first all the hernial rings, then the cæcum. If it be *distended*, obstruction must be in large intestine, and can be found by searching along the colon. If cæcum empty, search for an empty loop of small intestine, which can be followed up till the seat of trouble is reached.

If intestine sloughing, *enterectomy* (excision), and artificial anus or circular enterorraphy (suture).

Diseases of the Anus and Rectum.

Describe the varieties of congenital malformation of the anus and rectum.
1. Partial or complete *occlusion of the anus.* There is a membrane of varying thickness, bulging when the child cries or strains, and thin enough for the meconium to be detected.
2. *Imperforate anus.* The rectum terminates in a blind pouch, from half an inch to an inch from the surface; the normal position of the anus is occupied by dense tissue.
3. *Occlusion of the rectum.* A membranous septum is found from half an inch to an inch above the anal orifice.
4. *Imperforate rectum.* Rectum wanting. The colon terminates in a blind pouch in the iliac fossa.
5. *Malformation* with *abnormal opening in other parts.*

How do you treat congenital malformation of the anus and rectum?
Place the child in lithotomy position.
Incision in the middle line, over the natural position for the anus. Work backward toward the coccyx. The bowel being

found, open, and, if possible, suture to the external wound. Pass a bougie daily to prevent contraction. If, after dissecting to the depth of 1½ inches, no sign of bowel is perceived, do Littré's operation (left inguinal colotomy), making an artificial anus.

What are hemorrhoids?

Swellings about the margins of the anus due to a varicose condition of the bloodvessels. Hemorrhoids may be *external*, affecting the muco-cutaneous folds external to the sphincter, or *internal*, affecting the mucous membrane within the sphincter.

What are the causes of hemorrhoids?

Anything tending to increase the supply of blood to the rectum, or to impede its venous return. Instance, liver troubles, constipation, straining, occupations requiring much standing, sedentary life. They begin as dilations of the hemorrhoidal veins, and are followed by infiltration of surrounding tissues.

Describe external piles.

May be made up of dilated and thrombosed veins, *thrombotic*; may be due to swollen muco-cutaneous folds, *œdematous*; or may consist of permanently hypertrophied flaps or tags of skin, *cutaneous*. These occasion little trouble till, from cold, constipation, imprudent diet, or some other cause, they become inflamed, when they give rise to intolerable pain and itching, and exhibit all the local signs of an acute inflammation; this constitutes an "attack of piles."

Give the treatment of external piles.

Keep the bowels open by equal parts confection of senna and confection of black pepper, or a glass of Friedrichshall on rising in the morning; scrupulous cleanliness of the parts. Cocaine suppository (gr. ¼) for acute attacks.

Thrombotic. Apply a ten grain to the ounce calomel ointment at night and in the morning, after washing. If the parts become very painful, incise and turn out the clot.

Describe internal piles.

May be *open or bleeding, blind or not bleeding*.

1. *Capillary hemorrhoids.* Small, granular, bright red tumors, situated high in the bowel; really *arterial nœvi*.

2. *Arterial hemorrhoids.* Hard, vascular, glistening, slippery; may attain considerable dimensions. On scratching, bright red blood in jets. Large artery can be felt entering the upper part of each pile.

3. *Venous hemorrhoids.* Large, livid, prone to prolapse.

What are the symptoms of internal hemorrhoids?

Bleeding at stools. The blood is *bright red* and *coats the fœces. Protrusion.* An irregularly nodulated congested mass protrudes after defecation. It may become strangulated by the sphincter. *Constipation. Discomfort and heaviness* about the rectum. *Pain and fever,* if the piles are inflamed or strangulated.

Give the treatment for internal piles.

1. *Palliative.* Equal parts of senna and black pepper confection, a teaspoonful on rising. Coat the diseased area with ferri subsulph. ʒss, cosmoline ʒj. *Inflamed piles.* Laudanum and starch-water injections. Hot fomentations. *Cocaine suppositories* (gr. ¼). For strangulated piles, anæsthetize, and return within the sphincter.

2. *Operative.* Clear the lower bowel by laxatives and injection. *Lithotomy,* or the *lateral* position. (1) *Injection of carbolic acid.* Clamp the pile and inject ℞ v of a 20 per cent. glycerine and water carbolic solution into the *centre* of the pile. (2) *Ligature.* Paralyze the sphincter, draw down each pile, divide the *skin* about it, and encircle its base by a ligature; or *transfix* with a needle carrying a double thread, and tie each half separately. Insert an opium suppository and apply a T bandage with a compress of iodoform gauze. Open the bowel on the fifth day.

3. *Clamp and cautery.*

4. *Crushing.*

5. *Excision.*

Give the treatment for secondary hemorrhage after pile operations.

Cold injections. Insert rubber-bag and inflate with cold water. Pass in a full-sized drainage-tube and pack the rectum about it with styptic cotton or gauze (containing subsulphate).

Name the forms of prolapse of the rectum.

Partial, involving only mucous membrane.

Complete, involving all the tissues of the gut (really an invagination).

Name some of the causes of prolapse.

Relaxation. Undue straining. Irritation, such as that caused by ascarides, polypus, stone in bladder, phimosis.

Usually occurs in *children* or *aged people*.

Give the symptoms of prolapse.

A protrusion of a *soft, non-nodulated, non-pediculated*, smooth mass about the entire circumference of the anus, continuous with the mucous coating of the sphincter in the partial form.

Give the treatment of prolapse.

Reduce. Patient in knee-breast posture; bowel covered with oiled lint and pushed up. If strangulated, divide the sphincter. After reduction strap the nates together (plaster), keep bowels soluble, and let them be moved while the patient is in the recumbent or standing posture. The cold douche, or astringent injections are often serviceable.

Operative. 1. Take up longitudinal folds of mucous membrane in Smith's clamp, cut off with scissors, and cauterize pedicle (*clamp and cautery*). 2. Ligate portions of the mucous membrane. 3. Apply nitric acid to entire prolapsed surface, cover with carbolized oiled lint, and restore.

What is a fistula in ano?

An abnormal communication between the rectum and the surface.

Usual cause. Abscess.

Name the varieties of fistula in ano.

Complete, having a gut and a surface opening. The gut opening is usually just above the internal sphincter.

Incomplete or *blind*, having but one opening.

 a. *External*, opens on surface only.

 b. *Internal*, opens in bowel only.

What are the symptoms of fistula in ano?

1. *Discharge.* Thin pus, causing excoriations, and coating the fæces in the internal or blind variety.

2. *Local signs of inflammation,* which are subject to frequent exacerbations.

3. *Opening,* sometimes very small. On using a probe its end will be felt by the finger in the rectum, either passing into the bowel, or, if there be no internal opening, lying beneath the mucous membrane.

Give the treatment for fistula.

Operation. Pass a grooved director along the fistulous tract till its point is felt on the finger introduced into the bowel, hook it forward bringing it out through the anus, divide the structures thus raised upon the director *and all sinuses or pockets* communicating with the fistula. Do not divide the sphincter in more than one place. In women do not divide the sphincter, as it decussates with the vaginal fibres. Wipe out the wound with caustic potash, pack with lint saturated in carbolized oil, and allow the wound to heal from the bottom.

What is anal fissure?

Anal fissure is a *lineal ulcer* or *crack* usually *just within the anus.* Caused by constipation, and large hard passages.

Give the symptoms of anal fissure.

1. *Smarting pain* coming on after defecation, often *intense* and *radiating from rectum.* Smarting changed to an aching sensation, which may last for several hours.

2. *Fæces streaked with blood.*

How do you diagnose anal fissure?

Examination is *painful;* the sphincter and levator ani are spasmodically contracted. Two œdematous folds of mucous membrane are found, which being separated reveal the *ulcer.*

Give the treatment of anal fissure.

1. Keep the bowels loose (cascara sagrada gr. iij. at night, or Hunyadi Janos on rising), wash with soap and warm water after each passage, and apply ferri subsulph. (gr. x to ℥j cosmoline).

2. Anaesthetize the patient. Insert the thumbs into the anus, separate them till the ischial tuberosities are felt.

3. Local anaesthesia by cocaine (gr. xx to ℥j). Draw a bistoury longitudinally through the base of the ulcer from above downwards.

What other forms of ulceration occur about the anus and rectum?

Syphilitic, tubercular, senile (varicose).

Give the symptoms of ulcer of the rectum.

Tendency to morning diarrhœa. There is an urgent desire to open the bowels immediately on rising.

Pain, moderate. *Tenesmus,* relieved by evacuation.

Discharge. Mucus or muco-pus, at times containing also disintegrated blood.

Ulcerated surface is *seen* and *felt* on examination.

Give the treatment of ulceration of the anus and rectum.

Treat constitutional condition. Highly *nutritious diet, bowels soluble.* Night and morning, cleansing injections of warm boracic acid solution (ad lib.), or boroglyceride; at night starch water and laudanum gttxx by injection. In severer cases nitric acid directly to ulcer, applied through speculum.

Name the varieties of stricture of the rectum.

1. Fibrous. 2. Malignant.

What is the cause of simple (fibrous stricture)?

Inflammation or ulceration.

What are the symptoms of fibrous stricture?

1. *Constipation,* slowly increasing.
2. *Motions like pipe-stems,* or broken up into *scybala.*
3. *A sense of fullness after passages,* as though there were more to come.
4. *Diarrhœa, alternating with constipation,* or predominating. Constant desire to go to the closet, passage of very little solid, with yeasty liquid.
5. *Wind* which cannot be passed except in the closet, as it is accompanied by a liquid discharge.

6. *Excoriation* and inflammation of anus from discharge. Frequently fistula.

By examination the stricture can usually be felt.

Give the treatment for fibrous stricture of the rectum.

Gradual dilatation by means of bougies. Partial or complete division of the stricture. Inguinal colotomy. Excision of stricture.

Give symptoms and treatment of malignant stricture of the rectum.

Usually epithelioma; about half inch above anus. In addition to the signs of stricture, there is *intense pain* radiating from the seat of trouble, there is frequently free *bleeding*, and the *discharge* is profuse, offensive, watery, or often bloody, and becomes finally like coffee-grounds. Cancerous cachexia always develops. On examination, the abnormal growth is detected; indurated, nodulated, and, if the disease is advanced, with fungoid out-croppings over its surface, which break down under the examining finger, coating it with a blood-stained offensive muco-pus.

Treatment. Excision if the disease is strictly local, inguinal colotomy if there is systemic involvement.

Give the symptoms of impacted fæces.

Constipation, distension, pain, and very frequently a *spurious diarrhœa, i. e.,* a mucous semi-fæculent discharge, due to the irritation of the impacted mass.

Diagnosis by rectal examination.

Treatment. Break up the lower part of the mass with the finger or the handle of a wooden spoon, and wash away by means of copious injections.

Describe polyp of the rectum.

Two varieties. 1. *Fibrous.* Smooth surface, may reach large size. 2. *Adenoid.* Identical in structure with the mucous membrane; looks very much like a raspberry. Both usually pedunculated; occur in children.

Symptoms. Bleeding after stools, and prolapse.

Treatment. Ligate and remove.

Describe villous tumors of the rectum.

Practically a mass of non-pediculated adenoid polyps.

Symptoms. Hemorrhages, feeling of fulness in rectum, and thin, mucoid, glutinous discharge.

On examination a *lobulated, soft, velvety, movable* mass is found.

Treatment. Complete removal.

Describe pruritus ani.

Obstinate itching about the anus ; frequently depending on *local irritation* (as pediculi, threadworms, piles), or on gouty diathesis ; it may be without obvious cause.

Give treatment of pruritus ani.

Removal of cause, strict cleanliness, regularity in the motions from the bowels, exercise, Turkish baths. Suppositories of cocaine (gr. ¼) or iodoform (gr. v), morphine, carbolic acid, mercurial ointment. Alum and zinc sulphate, equal parts of each, fuse, powder, dissolve in ℨj aq. ; use as injection. A rectal plug may be worn at night.

VENEREAL DISEASES.

What is syphilis?
Syphilis is a constitutional disease, due to inoculation with specific virus.

What is the primary lesion of syphilis?
The chancre.

What is the period of primary incubation?
The time which intervenes between inoculation and the appearance of chancre. Rarely earlier than two weeks or later than five; average, three weeks.

What is the period of secondary incubation?
The time between the appearance of *chancre* and the development of *secondary symptoms*. Rarely before the first or after the third month succeeding the chancre.

When do the tertiary symptoms appear?
At a period varying from a few months to many years after the secondaries.

Describe chancre or primary sore.
Found commonly about the corona glandis, may appear anywhere. Contracted *directly*, by contact with chancre or secondaries (mucous patches), *indirectly* from articles used by syphilitics.

Appears as an indurated papule, which develops into an abrasion, tubercle, or ulcer.

What are the characteristics of the primary sore?
1. *Indurated base* and *thin, scanty secretions*.
2. *Inflammation slight* around the sore.
3. *Usually single, not autoinoculable*.
4. *Buboes* are *polyganglionic* and painless; rarely suppurate.
5. *Appears after an incubation period* and is *followed by secondaries*.

The Hunterian chancre is characterized by greater depth, freer discharge, and more marked induration.

The mixed chancre exhibits the peculiarities of both syphilitic and chancroidal inflammation.

Give the treatment of chancre.

Wash several times daily with black wash, and dust with calomel, subiodide of bismuth, iodol, or iodoform. *Do not begin mercury till the secondaries appear.*

Describe the secondary lesions of syphilis.

1. General enlargement of the lymphatic glands.
2. *Eruptions of the skin* and mucous membranes; at times, inflammation of the iris or periosteum, and falling of the hair.

Pathology. Congestion, infiltration, ulceration.

The development of secondaries is preceded by general malaise, fever, and anæmia, lasting a few days and disappearing on the appearance of *roseola* and *sore throat*.

The skin eruption may simulate the various forms of skin disease. It may be *erythematous* (s. roseola), *papular* (s. lichen), *vesicular* (s. herpes, eczema, and varicella), *bullous* (s. pemphigus), or *pustular* (s. ecthyma, acne, or variola).

Mucous membrane lesions.

Pathology, as in the skin, first congestion (syphilitic sore throat), then infiltration with maceration of the epithelium (mucous patches), finally ulcers.

What are the characteristics of syphilitic skin eruptions?

1. Absence of itching.
2. Symmetrical arrangement (on the two sides of the body).
3. Reddish-brown or coppery in color (raw ham).
4. Polymorphous (many kinds of eruption at the same time).
5. Therapeutic test (use of mercury).

Describe the mucous patch.

Synonyms. *Condyloma. Mucous tubercle.*

Pathology. A congested, infiltrated macule, the surface of which is, from its peculiar position (about the anus, on the scrotum, in the gluteal folds), continually moist, in consequence of which the epithelium becomes sodden.

Appearance. A somewhat elevated, flat macule, covered with a dirty whitish, offensive exudation.

Give the treatment of secondary syphilis.

Mercury; hydrarg. prot. iodid. gr. ¼ three times daily, guarding the bowels by opium. Increase the dose gradually till the patient exhibits the offensive breath or the beginning mouth tenderness of *ptyalism.* Then cut the daily quantity down one-half, and continue for eighteen months, unless new symptoms appear, when the dose may be temporarily increased. After eighteen months, add iodide of potassium, and continue for six months or a year.

Mercury may be given : 1. By the stomach. 2. By inunction. 3. By vaporization.

By inunction. Unguent. hydrarg. ℨss to ℨj at night; rubbed into the feet after they have been soaked in hot water. The same stockings must be worn night and day.

Mucous patches should be washed with black wash, and dusted with a powder made up of calomel one part, zinc-oxide two parts.

Sore throat is treated by astringent gargles.

Describe the tertiary lesions of syphilis.

Between the secondaries and tertiaries proper there are certain symptoms, called *reminders,* which sometimes appear. Among them are skin eruptions, enlargement of the testicle, choroiditis, ulceration of the tongue, disease of the arteries, and psoriasis of the palms.

Tertiary lesion of syphilis is the gumma. This has no tendency to spontaneous cure, and is characterized by the formation of masses of granulation cells, which commonly infiltrate the surrounding tissues, and break down in the centre.

A *gumma* may break down, leaving an ulcer, or may be *absorbed,* leaving *fibroid thickening* and *scarring* (syphilitic stricture of rectum and œsophagus, etc.). The gumma may attack the periosteum, causing nodes, caries, or necrosis; the cutaneous and mucous surface, causing ulcers on any part of the body.

These ulcers of tertiary syphilis are asymmetrical, and are *not contagious.*

Give the treatment of tertiary syphilis.

Mercury and *potassium iodide*, or *iodide of potassium* alone or combined with tonics. Commence with ten grains of potassium iodide three times a day, gradually increasing the dose till the desired effect is accomplished.

What are the characteristics of a tertiary ulcer?

Begins as a gumma or lump, which, when it breaks, exposes a gray slough, surrounded by granular tissues. The edges are rounded and sharply cut. Other signs of syphilis can be found. The affection yields to specific treatment.

Syphilitic leg ulcers usually involve the upper third.

What is congenital syphilis?

Syphilis transmitted to the fœtus through the spermatozoa of the father, or the ovum of the mother.

What are the characteristics of congenital syphilis?

Manifestations are rare before four to six weeks after birth; then there may be *secondaries*, as snuffles or coryza, macular or papular eruptions, mucous patches, ulcerations about the *mouth and lips* (rhagades), stomatitis, which, by its effect upon the dental sacs of the permanent teeth, causes the subsequent development of Hutchinson's teeth. After some years, tertiaries develop. These commonly take the form of interstitial keratitis, and gummatous developments.

Describe Hutchinson's teeth.

The upper permanent median incisors chiefly show this lesion, which consists in a dwarfing of the entire tooth, an extreme diminution in its free end, and a narrowing of the cutting edge, with a central notch or crescent.

Give the treatment of hereditary syphilis.

Upon the same lines as the acquired *secondaries*. Mercury best given by inunction, gr. x. unguent. hydrarg. being rubbed over the abdomen and covered by the belly-band every night. Stop mercury shortly after disappearance of symptoms. Prevent a non-infected woman from suckling the child.

Tertiaries. Mercury and iodide with tonics.

What is Colles's law?

A syphilitic child suckled by its mother will not infect her, though she be (*apparently*) free from venereal disease.

Chancroid.

What is a chancroid?

Chancroid is a local ulceration, caused by contact with the secretions of a similar ulcer.

What are the characteristics of chancroids?

1. *No period of incubation.* Appears in from three to five days; first as a papule, then a vesicle or pustule, very shortly an ulcer.
2. *Usually multiple.*
3. *Inflammatory in type*, with punched-out edges, irregular sloughing surface, and abundant discharge.
4. *Monoganglionic* and *unilateral* lymphatic involvement. May be *simple inflammatory enlargement*, or *virulent bubo* from direct absorption and suppuration.
5. *Autoinoculable.*
6. *Not indurated.*
7. *Not followed by secondaries.*

How may a chancroid be complicated?

Phagedenic ulceration. Characterized by very rapid and extensive sloughing.

Serpiginous ulceration. Characterized by slow but persistent extension.

Phimosis. Paraphimosis.

Give the treatment for chancroids.

1. Cauterize with hot iron, sulphuric or nitric acid. Dress with black wash or iodoform.
2. Cleanse thoroughly with acid. nitric. ℥ss, aq. f℥vij. Dust with iodoform, or zinc oxide one part, bismuth two parts.

Bubo. Try to abort by blisters, iodine *around* the inflamed area, or pressure by means of a salt or shot-bag. If it suppurates, open. If it is a *simple inflammatory* bubo, it quickly heals; if it is *chancroidal*, it has no tendency to heal, but must be thoroughly cauterized. After operation pack with iodoform gauze and dress antiseptically.

Phagadenic ulceration. Remove slough and *thoroughly cauterize.* Continuous warm bath is frequently curative. Internally, tonics, opium and iron, rich food, and alcoholic stimulants.

Serpiginous ulceration. Repeated applications of the actual cautery to the entire diseased surface, together with nourishing and stimulating internal treatment.

What is primary bubo or bubon d'emblée?

A simple adenitis resulting from mechanical irritation. It is seen at times, after coitus, when there is no taint of chancroid, gonorrhœa, or syphilis.

Gonorrhœa.

Describe the urethra.

Length, 8 to 9 inches.

Three portions. *Spongy, membranous,* and *prostatic.*

Spongy portion. 6 inches long from meatus to anterior layer of triangular ligament. *Meatus* narrowest portion of urethra. *Lacuna magna,* a large mucous follicle 1¼ inches from meatus on the upper surface of urethra; its opening is directed forward and may catch instruments. *Glandular* and *bulbous* parts of the spongy urethra somewhat dilated.

Membranous portion. ¾ inch long. From apex of prostate to beginning of spongy portion, between the two layers of the triangular ligament, 1 inch below pubic arch. *Except meatus,* the *narrowest part.* Embraced by compressor urethræ muscle.

Prostatic portion. 1¼ inches long. *Widest and most dilatable part;* passes through prostate near its upper surface.

What is gonorrhœa?

Gonorrhœa or *clap* is a *contagious* (probably specific) inflammation attacking mucous membranes, particularly those of the genito-urinary tract.

Cause. Direct contagion (genococcus). *Urethritis,* identical with gonorrhœa, is developed by contact with retained and foul discharges (leucorrhœa), or other irritants. *It begins* in the male usually in the fossa navicularis, and passes backward. In the female it begins in the vulva and vagina.

Name the clinical varieties of gonorrhœa.
1. Acute inflammatory (typical). 2. Subacute or catarrhal. 3. Irritative or abortive.

What are the stages of an acute attack?
First, or *increasing* stage. Second, or *stationary* stage. Third, or *subsiding* stage.

What are the first symptoms of gonorrhœa?
Usually, in three to five days, there is a tickling sensation at the meatus, which is changed to a burning at the next urination. On examination, the lips of the meatus are somewhat reddened and everted, and there is a slight muco-purulent discharge. In a very short time (twelve to twenty-four hours) the patient reaches the well-developed first stage.

Give the symptoms of the increasing stage.
1. Ardor urinæ. 2. Profuse purulent discharge. 3. Chordee (painful erections). 4. Frequent urination.

What are the complications of the first stage?
1. *Balanitis*, or inflammation extending over the glans penis.
2. *Balano-posthitis*. Inflammation of the mucous layer of the foreskin.
3. *Phimosis*, or inability to retract the foreskin, from œdematous swelling.
4. *Paraphimosis*. The retracted and swollen foreskin cannot be brought forward.

The first stage lasts about one week.

Give the symptoms and complications of the second stage.
The inflammation gradually *extends* backward. There is a continuance of the symptoms of the first stage, with possibility of the following complications:—

Follicular abscesses, appearing as small, round, tender tumors along the floor of the urethra. They may open either internally or externally.

Periurethral abscess. Favorite seat about the fossa navicularis and the anterior membranous portion of the urethra, where the disease is most persistent.

Lymphangitis. Dependent usually on retention of discharge beneath prepuce. Thick, tender, reddened cord-like line along dorsum of penis.

Bubo. One gland affected; may undergo resolution, or may suppurate.

Cowperitis. Characterized by very *intense throbbing pain.* Painful urination, especially at the end of the act (compressor urethræ m.), and the detection of the hard, inflamed glands by examination of the perineum.

Second stage lasts one or two weeks.

Give the symptoms and complications of the stage of subsidence.

Symptoms as of the other stages. They may be complicated by *epididymitis,* characterized by *pain* of an intense and sickening character passing along the cord to the loins, *swellings,* outlined at the back of the scrotum and considerable in extent, and *tenderness;* there is nearly always fever.

Describe subacute or catarrhal gonorrhœa.

Occurs usually in persons who have had previous attacks. Is characterized by *very free discharge,* with absence of other symptoms or complications. Yields rapidly to treatment, but does not *entirely* disappear, a drop or two of muco-pus being discharged daily.

What are the complications of subacute gonorrhœa?

Gonorrhœal rheumatism or urethral synovitis. Characterized by slight constitutional symptoms and a rapid development of synovitis in knee, ankle, wrist, or elbow.

Gonorrhœal ophthalmia (sclerotitis, iritis), or *conjunctivitis.*

Describe irritative or abortive gonorrhœa.

The symptoms are those of beginning acute gonorrhœa, *i. e.,* redness, pouting, and tingling or itching at the meatus, with a *very slight* mucous discharge. The disease does not advance beyond this point. These symptoms may persist for several days, then disappear. No complications, no sequelæ.

Give the treatment of gonorrhœa.

Rest in bed, if possible, on a diet of *skimmed milk*, giving plenty of bland liquids, such as Apollinaris water, soda water, etc. Keep the bowels open. To make the urine alkaline, and to act as a sedative, give—

    R. Tr. aconit. rad. gtt. xvj.
       Pot. brom. ʒiij.
       Infus. pareir. brav. fʒviij.
    S. fʒss in aq. every two hours.

For *ardor urinæ* give the above prescription. Immerse the penis in *hot water during urination*. Wrap the organ in cloths saturated with—

    Tr. aconit. rad.,
    Tr. opii,
    Alcohol, āā ʒj.
    Liq. plumb. subacetat. dil. fʒiij.

Chordee. Bromide of potassium till drowsiness is produced; a double dose on retiring, repeated during the night.

If the patient wakes with chordee, *camphor* gr. iij, *opium* gr. j, as a suppository or hypodermics of morphia (gr. ¼) injected into the perineum.

When the disease *has reached its height* and is *declining*, give capsules of cubebs and copaiba, ♏xx of each, every two hours.

Injections may now be used—

    Bismuth. subnit. ʒj.
    Glycerin, fʒij.
    Aq. ros. q. s. fʒiv.

Followed in a few days by—

    Zinc. sulph. gr. viij.
    Morph. sulph. gr. j.
    Aq. ros. fʒiv.

Gradually stop injections and internal medication.

What are the causes of chronic urethral discharge?

1. *Urethral catarrh.*
2. *Chronic gonorrhœa*, a localization of the disease, producing a granular and somewhat ulcerated surface.
3. *Stricture of urethra.* The usual cause of gleet.

VENEREAL DISEASES. 215

How can the nature of chronic urethral discharge be determined?

Urethral catarrh immediately follows gonorrhœa, and presents no symptoms beyond a thin, watery discharge.

Chronic gonorrhœa discharges creamy pus, is *greatly aggravated by any excess*. There is some *burning at urination*, and *at times*, chordee. It is generally found about the navicular fossa and the bulbo-membranous portion of the urethra. Examination by a *bulbous bougie* detects a *rough, tender spot*, and pus and blood may be brought away upon the shoulder of the instrument.

Gleet from stricture appears some time after subsidence of gonorrhœa. It is characterized by muco-purulent discharge, and *frequent urination, with imperfect cut off*. On passing a *bulbous bougie* narrowing is detected.

Give the treatment of chronic urethral discharge.

Urethral catarrh. Constitutional treatment, open air, nourishing diet, exercise, regular living, iodide of iron.

Chronic gonorrhœa. Locate the spot by means of the bulbous bougie. Apply, by means of the prostatic syringe, a one-quarter per cent. solution of nitrate of silver, increasing the strength if there is no pain; follow by astringents, zinc or copper.

Gleet. Gradual dilatation with steel sounds, passed twice weekly, till the urethra is of normal size (28 to 32, depending on the size of the penis).

Give the treatment for complications of gonorrhœa.

Balanitis. Wash carefully four times daily, and dust with iodol, iodoform, or a powder of bismuth and opium.

Balano-posthitis. Careful washing. If great swelling, envelop in lead water and laudanum.

Phimosis. Injections beneath the prepuce of soap and water, then water, finally lead water and laudanum; wrap the penis in cloths wet in lead water and laudanum. Incision or circumcision may be necessary.

Paraphimosis. Reduce by manipulating, or, covering the glans with lint, envelop it from before backward in an elastic band, slip a director under the constriction, remove the elastic wrapping, and reduction may be effected. *Incision* if other means fail.

Prostatitis, cystitis (see under these headings).

Epididymitis. Rest in bed, elevation of scrotum, application of evaporating lotions, abstraction of six or eight ounces of blood

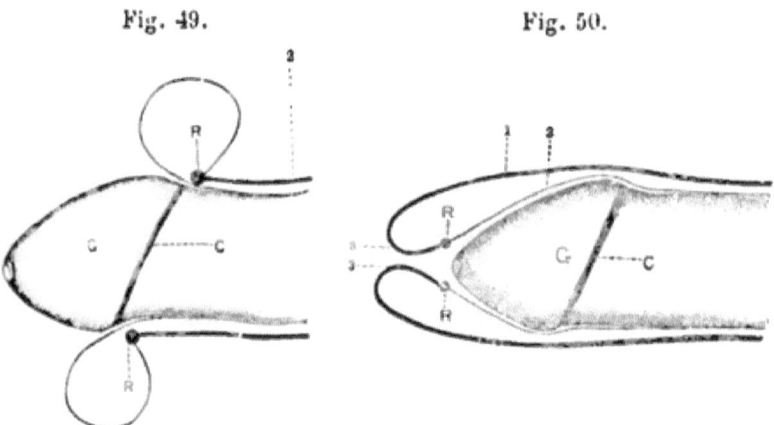

R. R. The constricting ring in paraphimosis.

R. R. The constricting ring in phimosis.

by leeches placed over the cord. Open the bowels, give morphia hypodermically, bromide of potassium and aconite internally. If swelling increases and pain is intense, puncture the tunica albuginea with a tenotome. When acute inflammatory symptoms begin to subside, strap the testicle.

Gonorrhœal rheumatism. Iodine and splint to the joint, together with firm pressure; very full doses of quinine (grains xl. daily), small doses of mercury, generous diet.

Give the treatment of gonorrhœa in the female.

Usual form, *vulvo-vaginitis*, may extend to the urethra, the womb, the Fallopian tubes (gonorrhœal salpingitis), and the ovaries.

Rest in bed, milk diet, free motion from the bowels, repeated daily washings with strong sod. bicarb. solutions, followed by thorough application of liq. argent. nit. grains lx to the ounce. General hot baths, or, in case of vaginitis, every two hours inject bicarbonate of soda solution, Oj, follow with aq., Oj, finally acetate of lead ʒiij (teaspoonful) in the pint of water. Keep the mucous surfaces apart by packing with absorbent cotton containing lead acetate.

Urethral Deformities.

Describe epispadia.
Epispadia, or deficiency of the urethral roof, may be *complete* or *partial*. Complete epispadia is usually associated with exstrophy of the bladder.

Treatment. Freshen the edges on either side of the urethral floor, and bring them together over a catheter by means of quill sutures; flaps may be transplanted.

Describe hypospadia.
Hypospadia, or deficiency of the urethral floor, may occur at the base of the frenum, or at the junction of the penis and scrotum.

Treatment. Restore the natural passage; freshen the edges of the abnormal opening, and close or cover by transplanted flaps.

Stricture of the Urethra.

What is stricture of the urethra?
True or organic stricture is permanent narrowing of the urethral canal at one or more places, due to disease, injury, or congenital defect. There are also *spasmodic* and *congestive* strictures.

What are the causes of stricture?
Gonorrhœa, traumatism, ulceration, and masturbation.

Give some varieties of organic urethral stricture.
In regard to cause: 1. Idiopathic. 2. Traumatic. 3. Inflammatory.

In regard to anatomical appearances—

1. *Bridle stricture.* A band of lymph, attached only by its ends, stretching across the urethra.

2. *Annular.* A circular constriction as though a string were tied about the urethra.

3. *Indurated annular.*

4. *Cartilaginous.*

In regard to the possibility of passing instruments strictures are classed as *permeable* and *impermeable*.

In regard to their behavior on manipulation, they may be *simple, irritable, contractile or recurring*.

What are the favorite seats of stricture?

1. Anterior part of the urethra. 2. Just in front of the membranous portion of the urethra. Strictures are never found in the prostatic portion of the urethra.

What are the consequences of an untreated stricture?

Hyperæmia and inflammation about the stricture. Dilation and thinning of the urethral walls behind. Hypersecretion and gleet. Ulceration may take place, followed by extravasation, abscesses, and fistulæ. From constant straining, bladder becomes thickened, hypertrophied, and sacculated. Urine is retained and ferments; cystitis may reach a high grade. The inflammation passes along the ureters, involves the pelves of the kidneys, and may cause death by suppurative pyelitis, or nephritis.

What are the symptoms of organic strictures of the urethra?

Gleety discharge, especially in the morning; *increased frequency of urination*, with some pain, *twisting, forking*, or diminution in the size of the stream. *Retention* may be the first and only sign. *Later symptoms* are due to involvement of other organs; hemorrhoids frequently result from constant straining.

How do you diagnose strictures?

By examination of the urethra with bulbous bougies. Commence with medium-sized bulbous bougie and increase the size till decided resistance is experienced; or if the first tried will not pass, diminish the size till one finally enters the bladder, marking on its stem the point where resistance begins; slowly withdraw from the bladder, marking again the point where resistance begins; this will give both the *calibre* and the *width*

STRICTURE OF THE URETHRA.

of the stricture. If the obstruction is more than six inches from the meatus, it is probably an enlarged prostate. The possibility of spasm or the catching of the bulb of the bougie in a lacuna or at the triangular ligament must be borne in mind.

What special points must be observed in passing a bougie or catheter?

1. See that the instrument is *clean*, *smooth*, and, if it is a catheter, *pervious*.
2. Warm and oil.
3. Place the patient on his back with thighs flexed.
4. Bear in mind the course of the urethra, keep the catheter in the middle line, stretch the penis forward and upward, and use no force.

What difficulties may occur in passing the catheter?

1. It may catch in a fold of mucous membrane or in a lacuna. Avoid by keeping the point on the *floor* of the urethra at first, then along its roof.
2. It may catch where the urethra enters the triangular ligament. Withdraw a little and keep the point of the instrument along the roof of the urethra.
3. It may make a new false passage, or enter one already made. Denoted by a sudden slipping of the instrument, *pain*, and detection of the point of the catheter outside of the urethra by rectal examination. The handle of the bougie is deflected from the middle line, no urine escapes, the point is not freely movable, and, if the false passage is recent, there will be *free bleeding*.

Fig. 51.

Bulbous bougie.

How do you treat false passage?

Withdraw the instrument at once, and make no further effort to pass it for one or two weeks. Infiltration of urine rarely takes place, the passage healing promptly.

What constitutional effects may follow the passage of an instrument?

Hæmaturia, due to reflex congestion, syncope, rigors, urethral fever, suppression of urine, pyæmia.

How may the danger from these sequelæ be lessened?

Pass instrument with the patient in the recumbent position; give 12 grains of quinine an hour before treating; inject ♏x to xx of a 1 per cent. solution of cocaine into the bulbous portion of the urethra by means of the prostatic syringe a few minutes before passing an instrument. Keep the patient in bed six to twenty-four hours after the instrument is used.

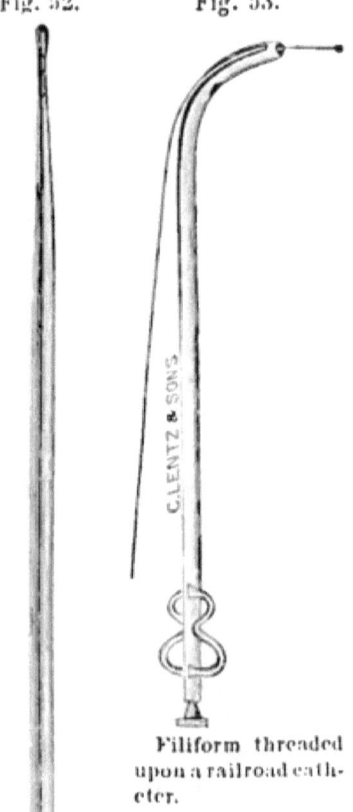

Fig. 52. Olive-pointed soft catheter.

Fig. 53. Filiform threaded upon a railroad catheter.

How do you treat strictures?

Strictures may be treated by —1. *Dilatation*. This may be *intermittent, continuous,* or *forcible* (splitting). 2. *Urethrotomy*, or cutting; either *internal* or *external*. 3. *Excision*. 4. *Electrolysis*.

How do you get through a tight stricture?

Try a small, soft, olive-pointed catheter or a small steel sound. That failing, electrolysis may succeed. Finally several filiforms should be passed into the urethra, and each manipulated in turn till one passes into the bladder; this may be threaded upon a railroad catheter and the latter forced through the stricture without fear of making a false passage.

Describe intermittent dilatation.

The calibre of the stricture having been determined, the largest flexible bougie which will pass through it is introduced, and allowed to remain in the urethra for four or five minutes before withdrawing. At the next attempt a larger instrument is used, till 28 to 30 French will readily pass in; three days should elapse between each dilation. This is the best and safest of all methods of treatment for the simple forms of stricture.

Describe continuous dilatation.

The patient is put to bed; a flexible catheter is passed through the stricture into the bladder, and allowed to remain for one or two days, when it is replaced by a larger one; continue in this way till the stricture is fully dilated.

Under what circumstances may continuous dilatation be employed?

Where there is great difficulty in passing an instrument, or where the stricture is irritable or contractile.

(The majority of surgeons condemn *rapid dilatation or splitting.*)

Describe internal urethrotomy.

By means of a guarded knife the stricture is cut *entirely through.* In tight strictures a guide or small instrument is passed, which can be threaded on the urethrotome, and the latter can then be made to cut its way inward without fear of its going astray. Pass a bulbous bougie to see that the stricture has been *completely* divided, in which case there is no fear of urinary extravasation. In four days pass a full-sized soft catheter.

What strictures are properly subject to internal urethrotomy?

Strictures in front of the scrotum, and contractile, irritable, and cartilaginous strictures.

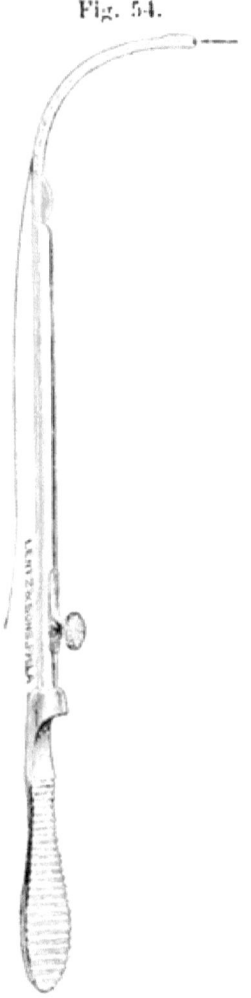

Fig. 54.

Railroad urethrotome. (White.)

Describe external perineal urethrotomy with a guide (Syme's method).

Lithotomy position. The groove of a Syme's staff is passed through the stricture till its shoulder is caught in the beginning of the narrowing. A 1½ inch incision is made in the median line of the perineum, the groove of the staff is found, the knife slipped into it *behind* the stricture, and the latter divided by pressing the cutting edge forward. A director is passed into the bladder, and a 14 (English) soft-rubber catheter passed per urethram. This catheter is not left in, but is passed every three or four days till the wound is healed.

Fig. 55.

Syme's staff.

What strictures call for external perineal urethrotomy with a guide?

Dense cartilaginous strictures, or irritable and contractile strictures when complicated by perineal fistulae.

How do you treat impermeable strictures?

By Wheelhouses's modification of perineal section.

By Cock's operation of perineal section, or tapping the urethra at the apex of the prostate.

What is Wheelhouses's modification of perineal section?

The urethra is opened half an inch *in front* of the stricture, when the latter can be exposed to view, entered by a probe, and divided. A broad director introduced into the bladder guides a flexible catheter passed through the meatus. The catheter is left in for three or four days. In this operation the Wheelhouses staff is used; this is practically a director, grooved to within half an inch of its end, and terminating in a blunt-hooked projection.

What are the indications for Wheelhouses's modification of perineal section?

Dense cartilaginous, or irritable and contractile strictures, which are impermeable.

Describe Cock's perineal section.

Lithotomy position. Left forefinger in rectum, the point applied to apex of prostate. Pass a long, straight knife, with its back towards the rectum, in the middle line beneath the bulb, so that it may enter the membranous portion of the urethra; a director is then introduced, and guided by it a soft catheter is passed into the bladder. The urethra is opened *behind* the stricture, the latter not being touched.

Indicated in case of impermeable stricture complicated by urinary retention, or in case of *urethral rupture.*

Describe rupture of the urethra.

Cause. Violence. May be torn partly or completely across.

Seats. Just in front of, or just behind the triangular ligament.

Give the symptoms of ruptured urethra.

Behind triangular ligament as in rupture of bladder. Inability to pass water. Blood and urine on catheterization. Infiltration behind symphysis.

In front of triangular ligament. Tumor in perineum; blood per urethram; inability to pass water.

How do you treat ruptured urethra?

Pass in a catheter. If there is any difficulty in introducing, do an external perineal urethrotomy, passing a catheter after two or three days, and at regular intervals afterwards. If urethra completely torn across, unite by catgut suture.

Describe urinary extravasation.

If extravasation takes place from the prostatic portion of the urethra, the symptoms and treatment are the same as for ruptured bladder. If from the membranous portion, there will be at first a hard lump in the perineum, as the anterior layer of the triangular ligament gives way, the extravasation will take the course common in all anterior extravasations, that is, into the scrotum and up upon the abdominal parietes, *not* descending upon the thighs (attachments of deep layer of superficial fascia). The symptoms are characteristic; if the patient has been suffering from retention, he may suddenly experience a sense of relief, followed shortly by burning pain in the perineum and inflammatory fever, which quickly becomes typhoid in type. There are redness, swelling, œdema, and early sloughing of the infiltrated area.

The *treatment* is perineal section, tapping the source of extravasation. Long incision should follow up the subcutaneous infiltration.

Diseases of the Prostate.

Name the surgical affections of the prostate gland.

Inflammation; may be acute, chronic, or complicated by abscess. *Hypertrophy. Atrophy. Tubercle. Malignant disease,* sarcoma in the young, carcinoma in the old.

Give the symptoms of acute inflammation of the prostate.

Usual cause—gonorrhœa or stricture.

There is *pain* at the neck of the bladder, *increased by defecation* and by *micturition,* especially towards the end of the act.

The *water is passed frequently.* On examination per rectum the prostate is felt as a *hot, tender enlargement.* There is *fever.*

Termination. Resolution, abscess, or chronic inflammation.

Give the treatment of acute prostatitis.

Open bowels freely. Render the urine bland by full doses of alkaline carbonates. Apply leeches to the perineum, followed by hot fomentations, poultices, and hot hip-baths. If there is retention, a catheter should be passed. If an abscess forms, open the perineum in the middle line.

Describe chronic prostatitis.

Causes. An acute attack, stricture, masturbation, gout.

It is characterized by *constant aching pain* in the perineum, aggravated by defecation and urination. There is a discharge.

like the white of an egg, appearing during defecation and at the beginning of urination. There is frequent urination with imperfect cut off, and cystitis.

Treatment. Avoidance of stimulants, sexual indulgence, or violent exercise. Bowels must be kept open. Tonics. Sea bathing. Fugitive blisters to perineum. By means of the prostatic syringe nitrate of silver, ℳ v of a two per cent. solution, applied to the diseased area.

Give the symptoms of enlarged prostate.

This is a disease of advanced life. It is characterized by *greatly increased frequency* of micturition, especially at night, by *loss of force in the stream,* with *difficulty* and *slowness* in starting it, by a *sense of fulness about the rectum.* Very frequently there are hemorrhoids from straining. Fermentation of retained urine with cystitis may follow. Finally, *retention with overflow,* or even *absolute retention* may result.

How do you diagnose an enlargement of the prostate?

The finger in the rectum will recognize most enlargements. In case there is projection of the middle lobe into the urethra a silver catheter will meet with an obstruction, *more than seven inches* from the meatus, which is only overcome by greatly depressing the handle of the instrument. An ordinary catheter may not be long enough to reach the bladder.

How do you treat chronic enlargement of the prostate?

Immediately after urination pass a soft catheter. If additional water can be drawn, it is proof that the obstruction prevents thorough emptying of the bladder. Give the patient a soft catheter, elbowed if the middle lobe is enlarged, and let him pass it every night on retiring. Commence this treatment *before* cystitis appears. *For more aggravated cases,* where the bladder is irritable and sacculated, the pain unbearable, the patient absolutely unable to pass water without a catheter, but suffering intensely each time the instrument is passed, *drain the bladder* by—1. Perineal section. 2. Suprapubic tapping and retention of canula. 3. Suprapubic incision with excision of a portion of the prostate.

15

What symptoms denote malignant disease of the prostate?

Pain, frequent urination, hemorrhage per urethram, shreds of growth in urine, rapid swelling of unequal consistency, glandular enlargements, cachexia.

Treatment. Palliative.

What is meant by bar at the neck of the bladder?

A ridge due to hypertrophy of the lateral lobes.

AFFECTIONS OF THE BLADDER.

Rupture of the Bladder.

Describe rupture of the bladder.

Causes. *Violence.* *Over-distension.* May be *intra-* or *extra-peritoneal.*

Symptoms. Pain and collapse, sense of something giving way, urgent desire to urinate without the power to do so, rapid development of inflammation or peritonitis. Catheter passed just inside the bladder draws *blood* only, or a small amount of bloody urine. If the patient has passed his urine immediately before the accident, a weak antiseptic solution (boracic acid) may be injected into the bladder. If there is a rupture, it cannot be again drawn off.

Treatment. Insertion of full-sized catheter and expectant, or *Suprapubic Cystotomy;* opening and washing out the peritoneal cavity if urine has been extravasated into it, closing the peritoneal rent, and inserting a drainage tube. After treatment, patient in lateral decubitus.

What tumors are found in the bladder?

Papilloma—most common benign tumor. *Mucous and fibrous polyps,* rare. *Sarcoma.* *Carcinoma,* epithelial or encephaloid. Tumors are usually situated on the trigone.

Give the symptoms of bladder tumor.

Hæmaturia, cystitis, pain, the *passage per urethram of fragments* of the growth.

Treatment. Benign growths may be removed by perineal or suprapubic operations.

Exstrophy of the Bladder.

What is exstrophy of the bladder?

Synonyms. Ectopion, extroversion.

Definition. Congenital absence of the anterior wall of the bladder, together with the corresponding portion of the abdominal wall. The posterior wall of the bladder projects as a round, vascular, red, ulcerated tumor, covered with mucous membrane, and exposing the orifices of the ureters.

Treatment consists in covering in the defect by deep and superficial flaps, which have their raw surfaces apposed, and offer both to the bladder wall and externally, skin surfaces.

This deformity is usually accompanied by epispadia.

Cystitis.

What are the causes of cystitis?

Cystitis, or inflammation of the bladder, may be *acute* or *chronic*.

Causes. Mechanical or chemical injury, or direct extension (gonorrhœa).

Give the symptoms of acute cystitis.

Pain, burning, may be very severe, located in the bladder and perineum. *Strangury*, a continual desire to void urine, which is spasmodically passed, a few drops at a time. *Tenderness*, well marked over the pubes, in the bladder region. *Urine*, scanty, highly colored, containing mucus, blood, and pus. *Fever*, directly proportionate to the grade of inflammation.

Give the treatment of acute cystitis.

Rest in bed. Diet of skimmed milk, with carbonated drinks. Bowels soluble. Leeches to perineum, or over pubes. *Hot hip-baths and hot poultices. Alkaline carbonates, hyoscyamus*, morphia and belladonna suppository. *If urine is ammoniacal*, the bladder must be washed out with antiseptic lotions (boracic acid gr. iv to ℥j); this failing, an external perineal urethrotomy with drainage of the bladder is indicated.

Describe chronic cystitis.

Symptoms as in acute, but milder. *Urine* often ammoniacal, very offensive, contains large quantities of ropy mucus and pus. Mucous membrane thickened, congested, ulcerated. Muscular coat thickened, fasciculated, giving the interior of the bladder a *ridged* appearance. Between the muscular ridges the mucous membrane may be forced outward by constant straining, forming sacculations, in which stones may form.

Treatment. Removal of cause, where possible. General hygiene. Milk diet, with free use of non-stimulating drinks. Triticum repens, uva ursi, copaiba, cubebs. When urine alkaline, benzoic acid. *Local washings.* Twice daily with boracic acid, or water hot as it can be borne. *In severe cases*, perineal cystotomy and drainage.

Atony and Paralysis of the Bladder.

What is atony of the bladder?

By atony is implied a loss of tone in the muscular fibres of the bladder, making it unable to expel its contents. The bladder is only partially emptied at each micturition; it gradually becomes more and more full till the condition known as retention with overflow is developed, simulating incontinence. The cause of atony is over-distension; it may arise in the course of low fever, from voluntary neglect, or from urethral obstruction.

Treatment. Catheter; cold douche to bladder and to lumbar spine.

Describe paralysis of the bladder.

Cause. Injury, or organic disease of nervous system, nervous exhaustion. If the neck of the bladder is affected, it causes *incontinence*. If the body of the bladder alone is involved, there will be retention.

Treatment. Catheter, tonics, strychnia, electricity.

Hæmaturia.

How can you determine the source of blood in the urine?

From the kidney. Blood is uniformly distributed through the urine. *From the bladder.* Comparatively clear urine is passed at first, followed by blood. *From the urethra.* Blood passes first, then urine.

What surgical affections may cause renal hemorrhage?

Contusion or jarring, congestion, inflammation, calculus, the uric acid diathesis, catheterism, malignant disease.

Give the causes of bladder hemorrhages.

Traumatism, calculus, inflammation, new growths.

Give the causes of urethral hemorrhages.

Injury, ulceration, calculus, erectile growths.

How are clots removed from the bladder?

By large suction catheter. By digesting the clots in the bladder. By urethrotomy or cystotomy.

Retention of Urine.

What are the causes of retention of urine?

Retention means simply inability to pass the urine from the bladder. Suppression means absence of the secretion.

The causes of retention are—

1. Impacted calculus or foreign body.
2. Alterations in the urethral walls, either *permanent*, as stricture and enlarged prostate, or *temporary*, as congestion and spasm.
3. Pressure from without the urethra, as in case of certain tumors.
4. Atony or paralysis of the bladder.

In retention due to stricture, the acute condition is generally brought about by an added spasm or congestion due to excesses or exposure.

After operations or injury, spasmodic retention is especially liable to occur.

Give the symptoms and signs of retention.

If the condition comes on slowly, the bladder may become enormously distended, with few local or constitutional signs other than those connected with urethral obstruction. Finally the urine dribbles away as fast as secreted, the bladder still remaining full. This constitutes the condition known as *retention with overflow*, and is diagnosed by outlining the full bladder by means of abdominal percussion, and by passing a catheter.

AFFECTIONS OF THE BLADDER. 231

In sudden and complete retention there is intense local pain, with rapid development of constitutional symptoms of a typhoid type.

The bladder, unless greatly stiffened and altered by previous inflammation, rises out of the pelvis, and can be readily detected by abdominal examination.

What are the consequences of retention?

Atony, cystitis, nephritis, rupture of either the bladder, or of the urethra *behind* a point of obstruction, or *retention with overflow*.

Give the treatment of retention of urine.

If the symptoms are urgent, immediate catheterization.

Retention due to spasmodic and congestive strictures.

Spasm and congestion are rarely sufficient in themselves to cause retention; they are usually associated with slight stricture or enlargement of prostate, and are brought on by exposure, debauch, or operation.

Treatment. Hot bath, and full dose of tr. opii (℥ xxx) by the rectum. If there is no spontaneous relief, pass a catheter. Open the bowels, and keep the urine unirritating.

Fig. 56.

Fig. 57.

Prostatic catheter.

Mercier's elbowed catheter.

Retention due to organic stricture. Attempt to pass a soft catheter or filiform, failing, give opium per rectum, and hot bath. If the urine is still not passed, anæsthetize and again attempt to pass an instrument; if unsuccessful, either incise, or make a suprapubic aspiration or puncture.

Retention due to hypertrophy of prostate. Usually due to congestion (congestive stricture), it is induced by debauch, etc. Try the elbowed catheter, the flexible catheter with stylet, which is somewhat withdrawn when the beak impinges on the prostate, the silver prostatic catheter. If passed with much difficulty, leave in. If bladder very full, draw off only a part of the urine (to avoid syncope and hemorrhage). Catheterization failing, do not try to relax, but immediately puncture, or aspirate above the pubes.

Retention due to atony and paralysis of the bladder (usually retention with overflow). Regular use of soft catheter.

Describe suprapubic tapping of the bladder.

Trocar and canula, full-sized, and with a marked curve, thrust through the abdominal wall just above the pubes and into the bladder *beneath* the peritoneal reflection. The trocar is withdrawn, and a rubber tube is passed through the canula and left in. In three or four days the tube is withdrawn, leaving a short sinus into the bladder, which may be kept open indefinitely.

When *temporary* relief is sought from retention, aspirate in the same region. Tapping may also be done through the pubes, through the perineum, through the rectum.

What are the varieties of incontinence of urine?

True incontinence. The urine dribbles away as fast as secreted. Due to either enlargement of the middle lobe of the prostate, or disease or injury involving the lumbar cord.

Nocturnal incontinence. Due to an abnormal reflex sensibility. Slight irritation, such as might be caused by worms or phimosis, causes micturition.

Treatment. For nocturnal incontinence, lateral decubitus, and regular emptying of the bladder once or twice during the night. Sponge baths night and morning, belladonna pushed to its physiological limit.

Stone in the Bladder.

What are the common varieties of calculus?
Uric acid. Oxalate of lime. Phosphatic salts. Among the less common varieties are the stones made up of urates, cystin, xanthin. Calculi are formed of concentric laminae, frequently made up of different materials (alternating calculi). They may be *single* or *multiple*, *free* or *encysted*, only one surface, in the latter case, being subject to deposit.

How may you infer the nature of a stone?
By an examination of the urinary sediment.

How may stone terminate?
In cystitis, pyelitis, nephritis.

Give the symptoms of stone in the bladder.
Pain. Chronic, aggravated by motion and jarring, felt across the loins and down the thighs; also an *acute* pain, referred to the end of the penis, and most intense towards the termination of micturition (the stone falls on the sensitive trigone and the bladder walls contract upon it).

Increased frequency of micturition during the day, or while the patient is moving about.

Hæmaturia. Slight, following micturition.

Sudden stoppage of the stream while micturating. *Cystitis.* Piles in adults. Elongated prepuce in boys (from pulling). Prolapse of rectum in children.

How do you diagnose cystic calculus?
Pass into the bladder a solid or hollow sound with a sharply curved bulbous beak. Insert a finger into the rectum. By manipulating the instrument, and turning it towards all portions of the bladder, the stone may be struck. The click of the sound against the calculus should be both heard and felt.

Under what circumstances may careful sounding fail to detect stone?
When the stone is encysted, or when it is coated with blood and mucus. If symptoms point to stone, sound repeatedly.

GENERAL CHARACTERS OF THE MORE COMMON TYPES OF CALCULI.

	SHAPE.	SURFACE.	FRACTURE.	COLOR.
1. Uric acid.	Ovoid or round.	Smooth or slightly warty. Susceptible of a fairly high polish.	Crystalline in proportion to its purity. Brittle, but hard.	Yellow to red, or reddish-brown.
2. Urates, chiefly of ammonia.	Ovoid.	Very smooth and earthy.	Earthy and inclined to crumble; homogeneous if it forms a whole calculus, but in general existing as laminæ.	Whitish-gray to fawn.
3. Oxalates.	Very irregular.	Tuberculated.	Crystalline and very hard.	Dark brown, even black.
4. Mixed phosphates.	Depends on that of nucleus.	Smooth and friable.	Chalky, soft, and breaking easily. Sometimes with many small crystals on the surface.	White or gray.

CHEMICAL REACTIONS OF THE MORE COMMON CALCULI.

1. Uric acid. } Insoluble in hydrochloric acid; soluble when warmed with alkalies. Disappear, or almost
2. Urates. } disappear, in blowpipe flame. With nitric acid and ammonia, give the murexide test. The latter may usually be distinguished from the former by giving off ammonia fumes when heated with a solution of caustic potash.
3. Oxalate of lime. { Soluble in hydrochloric acid. Do not disappear in blowpipe flame. } Insoluble in acetic acid. After being heated, effervesces on the addition of an acid.
4. Mixed phosphates. { Do not give the murexide test. Not soluble when warmed with alkalies. } Soluble in acetic acid. Fuses when heated.

How may vesical calculi be treated?

By *Litholysis* or solvent treatment, practically useless in treating bladder stones. *Lithotrity*, or crushing the stone in the bladder. *Litholapaxy*, or crushing and washing out at one sitting. *Lithotomy*, or cutting into the bladder and removing the stone.

What circumstances guide you in the choice of operation?

Litholapaxy, in adults as a rule.

Lithotomy is indicated.

1. *In children*, because the urethra is small, the bladder lies high, and lithotomy has given the best statistics.
2. *For large hard stones*, an oxalate stone with maximum diameter greater than one inch would indicate the cutting operation.
3. *In case of marked urethral stricture.*
4. *In aggravated cystitis or sacculation of bladder.* The incision, by providing drainage, would greatly ameliorate the bladder disease.
5. *In irritable urethra*, with tendency to urethral fever.

Mention some sequelæ of litholapaxy.

Rigors and fever, retention of urine, cystitis or prostatitis, hemorrhage, suppression of urine, phlebitis, and pyæmia. If death occurs, it is mostly due to the chronic kidney trouble.

Describe lithotomy.

May be *Perineal*. (1. Lateral. 2. Median. 3. Bilateral.) *Recto-vesical. Suprapubic.*

Usual operation. Lateral perineal. Prepare the patient by rest in bed, a laxative the night before, an injection the morning of operation. Anæsthetize, draw the urine, and inject six ounces of warm water. Pass into the bladder a full sized grooved staff and strike the stone. If it is not found, withdraw the staff and pass a sound. Failing to strike it with this, the operation should be postponed. If the stone is found, place the patient in *lithotomy position*, the soles of the feet being grasped in the palms of the hands, and secured by shackles or bandages in this position; bring the nates down over the end of the table, let an assistant hold the staff directly in the middle line hooked under the pubes, while the operator, seated facing the buttocks, passes the finger of his left hand into the rectum, and, with the knife in his right hand, makes an incision midway between the scrotum and anus, and just to the left of the middle line, downward and outward to below the anus and somewhat nearer the tuberosity of the ischium than to this opening. The incision divides skin, superficial fascia, external hemorrhoidal and superficial perineal vessels, and the corresponding nerves. Deepen the wound, cutting transversus perinei muscle and artery, the lower border of the triangular ligament, and, possibly, some fibres of the accelerator urinae. Search with the disinfected finger of the left hand for the staff, place the point of the knife in the groove, dividing the compressor urethrae and membranous portion of the urethra. Turn the blade somewhat toward the patient's left (the longest diameter of the prostate), and push it through the levator prostatae, and the gland itself, till it enters the bladder. Withdraw the knife, and twist the finger along the concave surface of the staff into the bladder. When the stone is touched and the staff taken out, pass the forceps along the finger; on withdrawing the latter, there will be a rush of water, which commonly carries the stone into the grasp of the instrument. See that the stone is grasped with its smallest diameter presenting, and exert traction in the axis of the pelvis. Encysted calculi must be removed by the finger and a scoop.

Apply no dressing; simply dust with iodoform. Urine comes through the lithotomy wound for two days, then from the urethra, owing to swelling; as inflammation subsides it again flows from the wound. Put the patient in bed, on his back, and with a rubber bed-pan to receive the urine.

What accidents may occur in lateral lithotomy?

Hemorrhage, from a wounded artery, or from the prostatic plexus.

Treatment. Tie the bleeding point. If that cannot be accomplished, hæmostatic forceps, or acupressure. *Venous hemorrhage* may take place some hours after the operation, the blood flowing into the bladder; in which case wash out all coagula, and check the hemorrhage by a petticoated tube packed with lint.

Other less common accidents are, *wound of rectum*, *wound of bladder*, and *tearing the urethra across*, the latter complication especially liable to occur in children. If the urethra is pushed off the staff, the operation must be abandoned.

Mention some causes of death after lithotomy.

Infiltration of urine, from opening of recto-vesical fascia; *diffuse inflammation*, from bruising; *hemorrhage*, *pyæmia*, *peritonitis*, *shock*, *cystitis*, *suppression of urine*.

Describe median lithotomy.

Pass a grooved staff as before. Feel the apex of the prostate with the finger in the rectum. Make an incision in the median line of the perineum, beginning ½ inch from the anus, and pass the point of the knife into the groove of the staff, nicking the apex of the prostate and dividing the membranous portion of the urethra.

What are the indications for median lithotomy?

Small stones, foreign bodies, exploratory incisions.

Describe suprapubic lithotomy.

This operation consists in opening the anterior wall of the bladder, below the peritoneal reflection.

Position. On the back, with the buttocks elevated. Inflate the rectum moderately, by means of a rubber bag distended with air or water. Draw the urine, and inject four to six ounces of boracic acid solution into the bladder. Incision through the linea alba, immediately above the symphysis. Tear through the fibrous and fatty tissues till the wall of the bladder is exposed. Draw the peritoneal reflection upward. Incise below its attachment. Enlarge, if necessary, by tearing, and extract the stone. The patient maintains the lateral decubitus, changing from one side to the other. This drains the bladder. A rubber air-cushion is arranged to receive the urine.

What are the indications for suprapubic lithotomy?

Large, hard stones, of a greater diameter than one-and-a-half inches.

This method of cystotomy is also advised in cases of tumor of the bladder. Many surgeons consider this operation as preferable in nearly all cases where the bladder has to be opened.

What are the symptoms of calculus impacted in the urethra?

Sudden stoppage of the stream, great pain, a drop or two of blood, and retention of urine.

Treatment. If possible, work it forward along the urethra, grasp and extract with urethral forceps. Stretch the skin over it and extract by a small incision, letting the wound granulate. If at the neck of the bladder, do a median lithotomy.

Hydrocele.

Name the varieties of hydrocele.

1. *Vaginal* hydrocele. This is the common variety; the serous effusion is in the tunica vaginalis testis. *Hydrocele* implies this form.

2. *Congenital hydrocele.* Arises from an imperfect closure of the communication between the peritoneal cavity and the tunica vaginalis.

3. *Encysted hydrocele of the testis* or *epididymis.* Really cystic growths from these structures. The fluid is often opalescent and contains spermatozoa.

4. *Encysted hydrocele of the cord.* A serous effusion into an unobliterated portion of the funicular part of the tunica vaginalis.

What are the symptoms of hydrocele?

A smooth, tense, elastic, fluctuating swelling in the scrotum; of *pyriform shape*, and *translucent*. The testicle lies *behind* it and near its lower part.

In congenital hydrocele the effusion can be slowly pressed back into the peritoneal cavity, to reappear when pressure is removed.

Give the treatment of hydrocele.

Palliative. Discutient remedies (especially in the congenital form), such as muriate of ammonia ℥ss to aq. ℥j, or weak solutions of iodine. Tapping and draining off the fluid by trocar and canula.

Radical. Tapping and injection of iodine. Incision, drainage, and antiseptic dressing. Excision of sac.

Describe tapping a hydrocele.

See that the trocar and canula are *clean*, and movable on each other. Determine the position of the testicle. Grasp the enlargement with the left hand, making its anterior portion tense. Thrust the trocar directly backward, turning it upward as soon as it has entered the sac. Evacuate the fluid, withdraw the canula, and close the wound with iodoform collodion.

If the hydrocele is to be *radically* cured, inject, after draining the fluid, tr. iodin. ʒij, and manipulate the scrotum so that the injection may come in contact with every portion of the sac walls. Withdraw the canula, and close the wound as before. Acute inflammation shortly follows, and the swelling may even exceed its original extent. It shortly subsides, obliterating the cavity by inflammatory adhesions.

Hæmatocele.

What is hæmatocele?

An effusion of blood into the tunica vaginalis testis. Strictly, the term includes effusion in connection with either testis or cord, as in case of hydrocele.

What are the causes of hæmatocele?

Traumatism, or spontaneous rupture of diseased bloodvessels.

How do you diagnose hæmatocele?

A smooth, tense, semifluctuating, pyriform swelling appears rather suddenly. It is opaque by transmitted light, gives to the exploring needle disorganized blood, and is often accompanied by considerable ecchymosis of the scrotum.

Give the treatment of hæmatocele.

If *recent*, rest in bed, elevation, and application of cold. If this fails, incise and evacuate.

Varicocele.

What is varicocele?
 A varicose condition of the pampiniform plexus.

Why is varicocele commonly found on the left side?
 1. The left spermatic vein is longer. 2. It opens into the renal vein at *right angles* to the blood-current. 3. It is crossed by the sigmoid flexure, and hence subject to pressure from fæcal accumulations.

What are the symptoms of varicocele?
 Dragging pain and discomfort, relieved by recumbency. *Considerable mental depression.* On examination there is found a soft, knotted, irregular, opaque, pyriform tumor, feeling like a bunch of earth-worms; it gives an impulse on coughing, and gradually disappears on lying down.

Give the treatment of varicocele.
 General hygiene, regular exercise, cold sponging, and local douches. The bowels should be regulated, and a suspensory bandage worn, with a ring through which a portion of the scrotum can be drawn.
 Radical. Subcutaneous ligation or acupressure. Excision.

Sarcocele.

Name the surgical affections of the testicle.
 Epididymitis and *orchitis*, acute or chronic. *Syphilitic, tubercular, cystic,* or *malignant* disease. All these enlargements may be accompanied by hydrocele.

What is sarcocele?
 A term applied to all solid enlargements of the testes, hence we have *simple, tubercular, malignant* sarcocele, etc.
 (For acute epididymitis see pages 213, 216. Acute orchitis has the same symptomatology and treatment.)

Describe simple sarcocele.
 Due to simple chronic orchitis. It is simply an overgrowth

of the connective tissue, following an acute attack of inflammation; forming a smooth, hard, non-sensitive enlargement. Testicular sensation may ultimately disappear. This indicates atrophy of the secreting tissues.

Treatment. Strap.

Describe syphilitic sarcocele.

Pathology. Either a diffused or localized induration (gumma). The testicle, at first smooth and globular, becomes *nodular*, of *stony hardness*, and *non-sensitive*. The tumor preserves its general ovoid outline.

Treatment. Strapping and constitutional medication.

Describe tubercular disease of the testicle.

The diagnostic points of *tubercular sarcocele* are: It occurs in the *young adult*, whose family history is frequently strumous, it is *indolent* and *slow* in development, the epididymis is first attacked, there is rarely hydrocele, the vas deferens is thickened, and the induration is prone to break down.

Treatment. Constitutional. Total ablation of diseased area. Castration if necessary.

Describe fibro-cystic disease of the testes.

Occurs in old men, and is a *gradual*, *painless*, *unilateral* enlargement, attended with absence of testicular sensation, and presenting no history of previous injury or inflammation.

Treatment. Castration.

Describe malignant disease of the testicle.

Sarcoma, most common, small round-celled. *Carcinoma*, usually encephaloid. The diagnosis from fibro-cystic disease is made by the exceeding rapidity of the growth, which involves the skin and ulcerates. All the signs of malignant disease are present.

16

DISEASES OF VEINS.

What is thrombosis?
A clot formed in a vessel during life.

What are the causes of venous thrombosis?
1. Inflammation, injury, or degeneration of the walls of a vein.
2. Alteration in the blood, blood stasis, or exhaustion.

What becomes of a thrombus?
It may organize, it may calcify, forming phleboliths, or it may undergo red or yellow (septic) softening.

What are the symptoms of thrombosis?
Œdema, and the detection of a tender, knotted, cord-like swelling in the course of a vein. There is pain on motion.

How do you treat thrombosis?
Rest and elevation. Mercury and belladonna ointment thickly applied, hot fomentations. Clear the bowels by a saline cathartic, give a simple but nourishing diet, and administer iron and quinine. Subsequently apply a pressure bandage, and use friction and massage.

What are the causes of phlebitis?
Traumatism, thrombosis, gout, micro-organisms.

What are the symptoms of phlebitis?
A dusky red line in the course of the vein, and the symptoms of thrombosis. Treatment as for thrombosis.

Describe suppurative phlebitis.
Cause. Septic micro-organisms.

Symptoms. As for phlebitis and thrombosis. Local inflammatory signs are more marked; there are frequently softening and suppuration in the course of the vein, and constitutional symptoms and metastatic abscesses indicate the development of pyæmia.

Prognosis. Unfavorable.

Treatment. Local disinfection and opening of abscesses; amputation, if the diagnosis can be made sufficiently early.

What is a varix?

A permanent dilatation of a vein. The vein is said to be varicose.

What are the causes of varicose veins?

Increased intravenous pressure from mechanical compression, from violent muscular contractions emptying the deep veins into the superficial, from long standing. Alteration in the vein walls.

What are the symptoms of varix?

Aching pains, and a sense of fulness after standing, together with the enlargement evident to the sight and touch. Muscular cramps are said to characterize deep varix.

How do you treat varicose veins?

Palliative. As much rest and elevation of the part as possible, the application of a rubber bandage or an elastic stocking, tonics, and laxatives.

Radical. Ligature and excision, or acupressure.

ANGIOMA.

Describe the different varieties of angiomata.

1. *Arterial varix.* A dilatation and lengthening of a single artery.

2. *Cirsoid aneurism.* A tumor composed of a number of dilated and tortuous arteries.

3. *Aneurism by anastomosis.* A dilatation and lengthening, involving the arteries, capillaries, and lesser veins.

4. *Capillary nævus.* A dilatation and tortuosity involving the capillaries.

5. *Venous nævus.* A tumor composed of a number of intercommunicating spaces lined with endothelium, into which the arteries empty, and from which the veins take their origin.

How do you treat angiomata?

Arterial varix, cirsoid aneurism, aneurism by anastomosis. Protect. If rapidly extending, excise, cutting free of the involved area, and tying each artery as it is cut. Ligation of the main artery of the part, or injection of perchloride of iron may also be tried.

Nævus. Very large superficial nævi (port-wine marks), and those which are neither increasing in size nor produce visible deformity, should not be treated. Under other circumstances *capillary nævi* may be removed by superficial cauterization, or incision, or escharotics lightly applied; *venous nævi* may be cured by *incision*, carried free of the diseased area; by *ligation*, the thread being placed subcutaneously, or in an incision made through the skin; by electrolysis, by coagulating injections.

ANEURISM.

What is an aneurism?
A blood tumor communicating with the interior of an artery.

Give the classification of aneurisms.
1. Traumatic (see p. 74).
    a. Diffused.
    b. Circumscribed.
    c. Arterio-venous.
2. Spontaneous.
    a. Tubular or fusiform.
    b. Sacculated.
    c. Dissecting.

The cirsoid aneurism and aneurism by anastomosis are, properly, varieties of spontaneous aneurism.

Describe spontaneous aneurism.
Tubular or fusiform. A circumscribed dilatation of the whole circumference of the artery. The sac consists of all three coats.
Sacculated. The dilatation involves a portion of the circumference only. The sac consists of the outer coat and of condensed areolar tissue. May be circumscribed or diffused.
Dissecting. The internal and a portion of the middle coat have yielded, the blood forcing its way between the layers of the middle coat.

What are the causes of spontaneous aneurism?
Predisposing. Atheroma, an embolus, leading to inflammatory softening.
Exciting. Blows, strains, or sudden violent exertion.

How may an aneurism terminate?
1. In spontaneous cure. 2. In death.

Spontaneous cure may be effected by, 1, gradual consolidation by deposit of laminated clot; 2, arterial occlusion above or below the sac by a fibrinous plug, or by the aneurism itself; 3, inflammation of the sac and consequent clotting of the contained blood; 4, suppuration and gangrene. Aneurism may cause death by pressure, by rupture and bleeding, by gangrene.

What are the diagnostic signs of aneurism?
A tumor in the course of an artery, diminished in size by

pressure of the main artery above, increased in size by pressure upon the artery below. Characterized by thrill, bruit, and expansile pulsation. The pulse in the artery below the aneurism is delayed in time, and more feeble than that of the opposite side of the body. There are various pressure effects, such as œdema, bony erosions, pain, muscular spasm, etc.

How do you treat aneurism?

1. *Medical treatment.* Absolute rest. Very restricted diet. Iodide of potassium.

2. *Surgical treatment.* (1.) *Pressure.* May be *direct,* upon the aneurismal sac, or *indirect,* upon the artery above or below. It may be *digital, instrumental,* or applied by an Esmarch's bandage. It may be so applied as to merely slow the blood-current producing laminated clots, or may completely stop the circulation (rapid pressure). (2.) *Flexion.* Usually combined with pressure. (3.) *Ligation.* The thread may be applied to the artery, 1, above the aneurism, and at some distance from it (Hunter's operation), 2, just above the aneurism (Anel's operation), 3, both above and below the aneurism (operation of Antyllus, or old operation), 4, just below the aneurism (Brasdor's operation), 5, to one or more of the main branches below the aneurism (Wardrop's operation). (4.) *Manipulation.* (5.) *Galvano-puncture.* (6.) *Injections.* (7.) *Introduction of foreign bodies.*

Describe the application of digital pressure to the cure of aneurism.

This, if it can be applied on the proximal side of the artery at some distance from the sac, is superior to other methods of pressure, since it is less painful, it is less liable to injure the soft parts, it does not obstruct venous circulation. This method can be combined with flexion and instrumental compression. Relays of assistants are necessary for its proper application. The pressure is made with the thumbs, the artery being controlled by the next assistant before the one pressing is relieved. A hand should be kept constantly on the sac to see that pulsation is prevented.

This method is not applicable to very large aneurisms accompanied by much œdema from venous obstruction, or aneurisms occurring in habitual drunkards or those of irritable disposition.

Describe Hunter's method of ligation.

The ligature is applied so high above the artery that a double collateral circulation is established, one around the thread, the other around the aneurism. The cure is effected by *diminishing the circulation*, and favoring the deposition of *laminated* clots in the aneurismal sac; these organize much more readily than the currant-jelly clots.

When the ligature is applied, pulsation can no longer be felt in the aneurism; after awhile a slight pulse is again perceptible; as the sac becomes occluded, this pulsation becomes more feeble, till it finally ceases permanently. After operation, the limb should be swathed in cotton, elevated, and kept warm.

What are the dangers of ligation?

Gangrene, secondary hemorrhage, suppuration and sloughing, recurrent pulsations.

What are the objections to ligation close to the aneurismal sac?

The artery is probably not healthy. The circulation is absolutely stopped, hence there is clotting in mass. The anatomical relations of the vessel are frequently altered by the tumor, making the operation difficult. The aneurismal sac is liable to injury during the operation.

How do you treat traumatic aneurisms?

Turn out the clots, and ligate above and below.

DISEASES OF THE LYMPHATICS.

Describe lymphangitis.

Definition. Inflammation of lymphatic vessels.

Causes. Septic absorption from a wound, or simple traumatism.

Symptoms. Irregularly placed erythematous patches, and red lines running to the nearest lymphatic glands, which are enlarged and tender. Chill followed by fever.

Treatment. Cleanse wounds and render aseptic. Promptly evacuate pus. Elevate and apply hot antiseptic fomentations. On subsidence of acute symptoms, apply belladonna and mercury ointment, together with pressure. Clear the bowels, give diaphoretics and diuretics.

Differential diagnosis. From phlebitis, by absence of knotted, corded feeling, and dusky redness in the course of veins; by the presence of glandular involvement.

Describe lymphadenitis.

Definition. Inflammation of lymphatic glands. May be acute or chronic.

Acute lymphadenitis is usually secondary to inflammation of soft parts. The symptoms are those of inflammation or abscess. The treatment consists in cleansing the source of trouble, the use of hot applications, prompt incision for pus, pressure, and applications of mercury and belladonna.

Chronic lymphadenitis. Common in strumous children, arises from slight irritation or without obvious cause. Glands of the neck frequently affected. Characterized by slow, painless, enlargements, which discharge curdy pus on breaking down, and leave indolent, undermined ulcers.

Treatment. Counter-irritation by iodine till signs of softening, then incise, curette, and dress antiseptically. Nourishing diet, fresh air, cod-liver oil, iodide of iron.

EFFECTS OF COLD.

How may death occur from cold?
From cerebral anæmia, caused by sudden and progressive chilling. *From cerebral congestion,* due to slow and continuous chilling. *From embolism,* due to sudden reheating.

Describe the local effects of cold.
Pernio or chilblain. Caused by sudden alterations in temperature. Characterized by swelling, congestion, vesication, and intense itching and burning. Frequent recurrence from slight causes.

Treatment. Restore circulation gradually by friction with snow, by the use of cold water. Apply a one per cent. solution of nitrate of silver, and wrap in raw cotton.

Frost-bite. Characterized by actual congelation of the part, which is brittle and of a tallowy whiteness; subsequently inflammation of a high grade appears, and may be followed by gangrene.

Treatment. Moderate the severity of reaction by rubbing with snow, continued cold irrigation, massage. If mortification appears, continue the use of cold as long as this process is inclined to spread. Amputate when the line of separation is formed.

FOREIGN BODY IN THE AIR-PASSAGES.

At what portions of the air-passages do foreign bodies become impacted?

Commonly in the larynx, or the right bronchus.

What are the symptoms of foreign body in the air-passages?

If impacted in the larynx. Asphyxia from spasm and obstruction: this may cause immediate death, or, the first spasm passing away, may be succeeded by an exhausting cough, a blood-stained mucous expectoration, and recurring spasmodic attacks.

If loose in the trachea. Recurring and violent attacks of spasmodic asphyxia from impact of the body against the rima glottidis, free secretion of a frothy mucus from the air-passages.

If impacted in a bronchus. Pain and whistling râles at the seat of lodgment, absence of respiratory sounds in the lung, abscess.

Treatment. If dyspnœa urgent, instant tracheotomy. If the foreign body is lodged in the larynx, an effort should be made to remove it by laryngeal forceps; failing in this perform laryngotomy and thyrotomy if necessary; let the patient wear a tracheal tube for twenty-four hours. If the foreign body is loose in the trachea, immediately tracheotomize, draw the wound open, invert the patient, and instruct him to cough. If the foreign body is lodged in a bronchus, endeavor to extract by means of wire or an instrument, passed through a tracheal opening.

What is bronchotomy?

Laryngotomy and tracheotomy, with their modifications. 1. Thyrotomy, opening through the thyroid cartilages. 2. Laryngotomy, opening through the crico-thyroid membrane. 3. Laryngo-tracheotomy, opening through crico-thyroid membrane, cricoid cartilage, and upper rings of the trachea. 4. Tracheotomy, opening through the rings of the trachea.

Under what circumstances is bronchotomy required?

Acute laryngitis, or œdema glottidis. Spasm. Emphysema. Foreign bodies in the air-passages, or gullet. Croup. Diphtheria. Polypi.

What structures lie in the middle line of the neck?

Thyro-hyoid membrane, thyroid cartilage, crico-thyroid membrane and arteries, cricoid cartilage, two or three tracheal rings, isthmus of the thyroid, trachea.

Describe laryngotomy.

Longitudinal skin incision, an inch-and-a-half long, is made over the thyroid cartilage, thyro-cricoid membrane, and cricoid cartilage; the crico-thyroid membrane is opened by a transverse cut.

Describe tracheotomy.

In the *high* operation the opening is made above the isthmus of the thyroid; in the *low* operation it is made below.

Incision for *high* operation, two and a half inches long, beginning at the upper border of the cricoid cartilage. Divide skin, superficial fascia, sterno-hyoid and sterno-thyroid inter-muscular fascia, and loose cellular tissue. Avoid anterior jugular veins and their communicating branch, inferior thyroid vein, and middle thyroid artery, if present. Draw the trachea forward with a tenaculum, incise, cutting from below upward, and pass in the tracheal tube. Check all bleeding before opening the larynx, except when death from asphyxia is imminent, or when the bleeding is due to intense venous engorgement.

After-treatment should be conducted in a warm, moist atmosphere; the opening of the tracheal tube should be protected by moist gauze, and a physician or nurse should be constantly present to clean the inner tube when it becomes filled. When the breathing becomes hissing, and the epigastrium and intercostal spaces are sucked in during inspiration, the tube is dangerously clogged. Bronchitis, pneumonia, or the disease which necessitates the operation, are the common causes of death after this operation.

Affections of the Œsophagus.

Where are the narrowest portions of the œsophagus?

At its commencement (the lower border of the cricoid cartilage), and as it passes through the diaphragm.

What are the symptoms of foreign body in the œsophagus?

Pain, difficulty in swallowing, and frequently, asphyxia from spasm or direct pressure.

How do you treat foreign body in the œsophagus?

If suffocation threatens, tracheotomize at once. Under other circumstances, endeavor to extract by forceps, or by the swivel or horsehair probang. If the body is of such a nature that it can be digested, or passed by the bowel, push it into the stomach. If the body is irregular and tightly lodged, perform œsophagotomy.

Describe stricture of the œsophagus.

1. *Spasmodic.* Occurs in young hysterical women. Gives trouble only at times. Under ether, a bougie is passed without difficulty.

2. *Fibrous.* Due to contractions following traumatism or syphilis.

3. *Malignant.* Generally epitheliomatous. Occurs opposite cricoid cartilage, tracheal bifurcation, or at cardiac end of stomach.

Symptoms of fibrous or malignant stricture are, increasing difficulty in swallowing, first solids then liquids giving trouble. A feeling of obstruction referred to the top of the sternum, regurgitation of swallowed food, progressive wasting. Finally the diagnosis is made by passage of bougies (after excluding aneurism, which has been burst by this procedure).

Treatment. Dilatation or internal œsophagotomy for fibrous strictures. Œsophagotomy (establishment of a fistulous opening into the œsophagus), or gastrostomy for malignant strictures.

SURGICAL AFFECTIONS OF THE BREAST.

In what situation may abscesses of the breast occur?
Supra-mammary, superficial to the gland. *Intra-mammary*, within the gland. *Post-mammary*, behind the gland.

Give the treatment of mammary abscess.
Early and free incision in a direction radiating from the nipple, drainage, and pressure by means of bandages or concentric strapping.

What is Paget's disease of the nipple?
An inflammatory condition of the nipple and areola which frequently precedes the development of cancer.

What tumors are most frequently found in the breast?
Scirrhus, fibroma, sarcoma.

Give the differential diagnosis between scirrhus and non-malignant breast tumors.

Scirrhus.	*Non-malignant tumors.*
Occurs after the fortieth year.	Occurs before the fortieth year.
Very hard, nodulated, shortly becomes fixed.	Nodulated, moderately hard, elastic, movable.
Skin infiltrated and adherent.	Skin free and movable.
Nipple retracted, superficial veins dilated, lancinating pain.	None of these signs present.
Lymphatic involvement, rapid growth, quick recurrence, cachexia.	

CLUB-FOOT.

Describe the common forms of club-foot.

1. *Talipes varus.* The sole of the foot looks inward. This is the commonest congenital form (usually equino-varus); when it affects both feet it is frequently associated with spina bifida. *Cause.* Contraction of tibialis anticus and posticus, muscles of the calf, and the plantar fascia. *Treatment.* Division of all resisting tissues.

2. *Talipes equinus.* The heel is raised. *Cause.* Contraction of gastrocnemius and soleus, or paralysis of the opposing muscles. *Treatment.* Division of tendo Achillis.

3. *Talipes valgus.* The foot is everted. Caused by long-continued standing, or anything tending to obliterate the plantar arch; the peronei muscles subsequently contract. *Treatment.* Friction, support to the arch of the foot, and section of peronei tendons, if necessary.

4. *Talipes calcaneus.* The toes are raised by the extensors. *Causes.* Contraction of the anterior muscles, or paralysis of those of the calf. *Treatment.* Section of the tibialis anticus, extensor longus pollicis, extensor longus digitorum, peroneus tertius.

There may be a combination of distortions, constituting equino-varus, calcaneo-varus, etc.

HARE-LIP AND CLEFT PALATE.

What is hare-lip?

A congenital deformity, characterized by a fissure or fissures on the upper lip, due to arrested development. Hare-lip is *single* when one side is involved, *double* when it appears on both sides. It is frequently associated with *cleft palate*.

The treatment consists in closing the fissure, by freshening the edges and bringing them together with hare-lip pins, or by performing a plastic operation, sacrificing none of the tissues.

What is cleft palate?

A congenital cleft in the median line of the palate; it may be confined to the uvula, the soft palate, or involve the entire roof of the mouth.

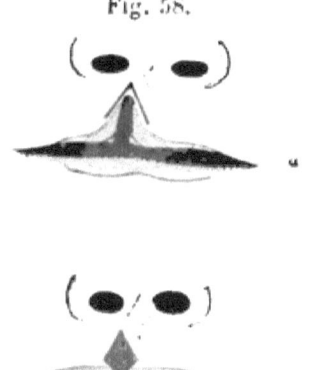

Fig. 58.

Operation for hare-lip.

Staphylorraphy indicates the operation for the closure by suture of the soft palate. The method of closing the fissure by a transparent flap from the pharynx is termed *staphyloplasty*. The flap operation for the closure of clefts in the hard palate is termed *uranoplasty*.

DISEASES OF BURSÆ AND TENDONS.

Bursitis.

Describe bursitis.

Bursitis is characterized by pain, fever, and the rapid development of a fluctuating swelling. The bursa patellæ is commonly involved, constituting, in the chronic form, "housemaid's knee." This swelling is diagnosed from intra-articular effusions by the fact that it is *above the bone*. Inflammation of the bursa over the olecranon constitutes "miner's elbow." "Weaver's bottom" is an inflammation of the bursa over the tuber ischii.

Treatment. Leeches, evaporating lotions, counter-irritation, and splinting. If suppuration, free incision.

How do you treat dropsy of a bursa?

This condition is usually due to subacute inflammation, or long-continued pressure. It may, at times, be resolved by counter-irritants, more commonly it will require incision and scraping.

What is a bunion?

A bursal enlargement occurring in the foot. It is usually placed at the side of the metatarsal joint of the great toe.

What is tenosynovitis?

Inflammation of tendons and their sheaths; due to traumatism, gout, or rheumatism. *Characterized* by a puffy swelling along the tendon, and fine crackling crepitation. *Treated* by iodine or blisters.

What is a ganglion?

A cyst formed in connection with the sheath of a tendon. The *simple ganglion* is developed *on* the synovial sheath. The *compound ganglion* consists of a dilatation which commonly involves the sheaths of several tendons. Ganglion occurs upon the extensor tendons at the back of the wrist, and in front of the ankle. It can be felt as a round, tense, fluctuating, freely movable tumor, sometimes giving considerable pain on motion, and always causing some loss of power.

Treatment. Subcutaneous rupture, either by force or by the tenotome. Incision and curetting.

What is paronychia?

Synonyms. Whitlow. Felon. Panaris.

Definition. An acute septic inflammation, involving the sheath of the tendon, the tissues superficial to it, or the periosteum, or all these structures. Always due to a septic wound. Characterized by intense pain, rapid disorganization, and tendency to spread along the course of the tendon. Treated by early, free incision, scraping, and thorough disinfection.

Onychia.

What is onychia?

Inflammation of the matrix of the nails.

May be *simple onychia* or "run around," due to injury, and attended by suppuration and loosening of the nail. *Treated* by wet boric acid dressing.

Malignant onychia, due to injury and profound constitutional depression; characterized by fungous ulcerations, showing no tendency to heal. *Treated* by trimming the nail, and applying powdered nitrate of lead to the granulations.

What is ingrowing toe nail?

An ulceration, caused by tight shoes pressing the soft part of the toe against the edge of the toe nail. Remedied by wearing loose shoes, packing absorbent cotton and iodoform between the soft parts and the nail, or by avulsing the nail.

17

ANÆSTHETICS.

What substances are used to produce anæsthesia?

General anæsthesia is induced by nitrous oxide, chloroform, or ether. Local anæsthesia is induced by cocaine or freezing.

Which is the safest general anæsthetic?

Nitrous oxide for brief operations (one minute), ether for manipulations requiring more time.

What is the danger in chloroform inhalation?

Cardiac syncope. It may attack the robust and apparently healthy. Particularly liable to occur when operations about the anus are begun before complete anæsthesia.

How do you prepare patients for the administration of anæsthetics?

Give no food for six hours before the time of administration. Examine the urine, and carefully auscult the lungs and heart. Half an hour before the administration of the anæsthetic give a full dose of whiskey or wine. See that there are no artificial teeth or foreign bodies in the mouth. Loosen the clothing about the neck and chest. In drunkards the anæsthetic should be preceded by a quarter of a grain of morphia.

How do you administer ether?

Use a folded towel, or one of the many inhalers. The recumbent position should be enforced. Protect the eyes by a folded towel. Let the vapor be very dilute for the first few inhalations, increasing the strength as the patient loses consciousness. Persistent cough is most quickly overcome by pushing the ether. *Watch the respiration and pulse.* When the pulse is slow and full, the respirations deep and snoring, the reflex irritability abolished, and the patient totally relaxed, the anæsthesia is carried to the limit of safety.

What accidents may occur during the administration of ether?

In the first stage there may be *respiratory forgetfulness*, or a cessation of breathing efforts, though consciousness is still preserved. Corrected by sudden pressure, or a dash of ether over the epigastrium.

In the *third stage* mucus may collect in the throat to such an extent as to embarrass respiration; it should be mopped out by sponges tied to sticks. If there is vomiting, the head should be turned to the side. If the air does not enter the lungs freely, the lower jaw should be pushed forward by the fingers placed beneath the ramus.

There may be threatened *asphyxia*, from excess of ether, dropping back of the tongue, or closure of the glottis. Denoted by irregular pulse, laryngeal stertor, blue surface, absence of respiratory movements. Immediately push the angles of the jaw forward or draw the tongue out of the mouth, practise artificial respirations, dash ether over the epigastrium, raise the foot of the bed or table, and apply the two poles of a battery, one to the right phrenic in the neck and the other to the sixth intercostal space, closing the circuit during the inspiratory movement of artificial respiration. Tracheotomy may be performed and the lungs inflated directly.

What precautions are taken during the administration of ether?

Lights, if near, should always be held *above* the level of the ether. A third person should always be present when women are etherized.

What are the indications for allowing the patient more air?

A feeble infrequent pulse. Lividity of the surface. Laryngeal stertor. Pallor and tonic spasm. A pupil fixed in dilatation (always a sign of great danger).

Under what circumstances is chloroform preferred to ether?

When there is emphysema of the lungs, bronchitis, kidney disease, or vascular degeneration. In infants. In operations about the mouth.

How do you administer chloroform?

The vapor must not be stronger than four parts to the hundred of air. Pour a few drops upon a piece of lint or a towel and hold it a short distance from the mouth and nose. Watch the pulse most carefully.

How do you treat syncope in chloroform narcosis?

Pull the tongue forward. Raise the foot of the table high up. Dash cold water over the face and chest. Begin artificial respiration immediately.

Should you give ether in shock?

As ether directly lowers the temperature, it should not be given when shock is marked. After restoration of temperature and full drugging with whiskey and opium, a minimum quantity will be required, and may be cautiously administered.

LIGATION OF ARTERIES.

Under what circumstances is an artery ligated in its continuity?
1. In the treatment of aneurism.
2. In the checking of bleeding, under certain circumstances.
3. In the treatment of inflammation.

What instruments are required for the operation?
Scalpel, dissecting and artery forceps, blunt hooks, retractors, grooved director, aneurism needle, ligature, needles, and dressings. All should be arranged in trays and covered with carbolic solution 1:20; which is diluted up to 1:40, when the operation is begun.

Describe the ligatures and dressings.
Ligature of antiseptic, prepared cat-gut. After operation, the wound, if small, is closed without drainage; if large, it is drained by means of rubber tubes, horsehair, or strands of cat-gut. Its edges are closely approximated, and the whole covered in by a careful antiseptic dressing.

What precautions are taken in performing the operation?
1. Begin and end the superficial cut with the knife-blade vertical to the surface, thus avoiding "heeling."
2. Divide the deep fascia to the full extent of the superficial cut. Open the sheath by cutting *toward the dissecting forceps*, in which a portion of its periphery is pinched up. The incision is subsequently enlarged by the director. Avoid forcible tearing or wide separation of the artery from its sheath. Pass the aneurism needle *from* the side where the most important and vulnerable structures are placed. Before tying, compress the artery and feel for pulsation below, to be sure that the circulation is controlled.

In securing the ligature, make more tension upon the first than upon the second knot.

What complications may arise in the after-treatment of ligation?
Gangrene, hemorrhage, return of pulsation in aneurism.

Describe the after-treatment of ligation.

Elevate the limb and surround it with a *thick layer of wool.* Keep at absolute rest. Light, nutritious diet. Strict quiet, both mental and physical.

Describe the triangles of the neck.

Anterior triangle. In front, the middle line. Behind, the sterno-cleido-mastoid. Above, the base of the lower jaw, and a line from its angle to the mastoid process. Apex, at the sternum. Subdivided into three smaller triangles by the digastric above, and the anterior belly of the omo-hyoid below, named from below up, the *inferior carotid,* the *superior carotid,* and the *submaxillary.*

Inferior carotid triangle. In front, middle line. Behind, sterno-mastoid. Above, anterior belly of omo-hyoid.

Superior carotid triangle. Behind, sterno-mastoid. Below, anterior belly of omo-hyoid. Above, posterior belly of digastric.

Submaxillary triangle. Above, body of jaw, parotid gland, and mastoid process. Below, posterior belly of digastric, and stylo-hyoid. In front, median line.

Posterior triangle. In front, sterno-mastoid. Behind, trapezius. Below, clavicle. Apex, at occiput. Divided by the posterior belly of the omo-hyoid into an upper or *occipital,* and a lower or *subclavian triangle.*

Occipital triangle. In front, sterno-mastoid. Behind, trapezius. Below, omo-hyoid.

Subclavian triangle. Above, posterior belly of omo-hyoid. Below clavicle. In front, sterno-mastoid.

Common carotid. *Origin*—right, from the innominate, behind the sterno-clavicular articulation; left, from the arch of the aorta, more deeply placed. *Extent*—from behind the sterno-clavicular articulation to the upper margin of the thyroid cartilage. The carotid artery lies in the same sheath with the internal jugular vein and the pneumogastric nerve, each of these structures being separated from the other by fibrous septa, and having a distinct compartment. The sheath rests upon the longus colli, and, in the upper part of its course, the rectus capitis anticus muscles, and is crossed at the level of the cricoid cartilage by the omo-hyoid muscle.

LIGATION OF ARTERIES. 263

Line. From the sterno-clavicular articulation to a point midway between the angle of the jaw and the mastoid process. *Superficial guide.*—anterior border of sterno-cleido-mastoid.

Relations. Anterior. Skin, superficial fascia, platysma, deep fascia, sterno-hyoid, sterno-thyroid, sterno-mastoid muscles; su-

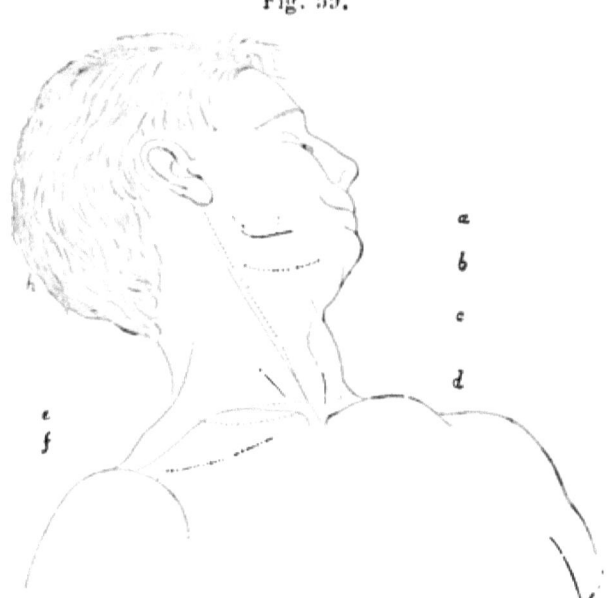

Fig. 59.

Lines of incision for carotid, facial, lingual, subclavian, and axillary arteries.

perior and middle thyroid, and anterior jugular veins; descendens noni and communicans noni nerves. *Posterior.*—Longus colli and rectus capitis anticus muscles; sympathetic, recurrent laryngeal nerves; inferior thyroid artery. *Internal.*—Trachea, œsophagus, larynx, pharynx, recurrent laryngeal nerve, and inferior thyroid artery. *External.*—Internal jugular vein, inferior thyroid artery. On the left side the internal jugular vein is somewhat anterior to the artery.

Collateral circulation. Inferior with superior thyroids, ascending branch of transversalis colli with princeps cervicis, terminal branches of internal and external carotids on the two sides.

Operation, above the omo-hyoid. Patient supine with a pillow under the shoulders and neck, head extended, face turned towards sound side. *Incision*, three inches, along the anterior border of the sterno-cleido-mastoid muscle, and with its centre on a level with the cricoid cartilage. Divide *skin, superficial fascia, platysma, deep fascia.* With retractors draw aside the sterno-mastoid. Expose the omo-hyoid by cutting through a dense fascia covering it and the sheath of the vessels, carefully avoiding the venous plexus formed by the superior thyroid with its communications from the lingual, facial, anterior and external jugular. The sheath of the artery is found bisecting the angle made by the anterior belly of the omo-hyoid and the anterior border of the sterno-mastoid. Open the inner compartment of the sheath, avoiding descendens and communicans noni nerves, and pass the ligature from without inward.

External carotid. A branch of the common carotid, given off at upper border of thyroid cartilage. It extends from the superior border of thyroid cartilage, to neck of condyle of lower jaw.

Chief relations. Anterior. Hypoglossal nerve, lingual and facial veins, digastric muscle. *Posterior.* Superior laryngeal and glosso-pharyngeal nerves. *Internal.* Hyoid bone and pharynx. *External.* Internal carotid artery and internal jugular vein.

Collateral circulation. Lingual, superior thyroid, occipital, and the same of the opposite side.

Operation. Incision midway between angle of jaw and anterior border of sterno-cleido-mastoid muscle, carried down three-eighths of an inch in front of the latter to one-half inch below upper border of thyroid cartilage. Divide skin, superficial fascia, and platysma at once. Slit up the deep fascia spreading from the anterior border of the sterno-cleido-mastoid, avoiding the external jugular, temporal, and facial veins. By blunt dissection the parotid gland and the posterior belly of the digastric are exposed; the latter is drawn upward with blunt hooks, when the external carotid is found, crossed by the hypoglossal nerve, with the superior laryngeal nerve lying beneath.

Pass the needle from without inward.

Lingual. Is given off from the external carotid between the superior thyroid and facial.

In the *first part* of its course, from its origin to the posterior border of the hyoglossus, it passes obliquely up and in to the great cornu of the hyoid bone, and is covered simply by skin, fasciæ, platysma, and veins, resting on the middle constrictor. In the *second part* of its course, beneath the hyoglossus muscle, it runs parallel with the great horn of the hyoid, then ascends to the tongue. It is crossed here by the posterior belly of the digastric and the stylo-hyoid muscles, and is covered by the hyoglossus muscle.

Chief relations. Anterior. Hyoglossus muscle. *Posterior.* Middle constriction of pharynx, and genio-hyoglossus muscle. *Above.* Hypoglossal nerve. *Below.* Tendon of digastric, and great horn of hyoid bone.

Point of election. Second part of artery, lying beneath hyoglossus.

Operation. Incision three inches; begin a little below and internal to the symphysis menti, convex downward to the great horn of the hyoid, and outward to the inner border of the sterno-mastoid. The three outer layers being divided the submaxillary gland is reached, lying in the deep fascia; the latter is divided and the gland turned up exposing the tendon of the digastric, and the hypoglossal nerve above; the nerve is dissected up and retracted exposing the hyoglossus muscle, which, when divided upon a director, enables the operator to pass the ligature about the artery from above downwards. Superficial guide, great horn of hyoid. Deep guide, nerve and tendon.

Facial arises from external carotid, a little above the lingual, passes beneath the posterior belly of the digastric and stylo-hyoid muscles and hypoglossal nerve, winds through a groove in the posterior and upper border of the submaxillary gland, and crosses the lower jaw in a slight depression just in front of the insertion of the masseter muscle. Here is the point of election; the artery is covered at this point by skin fascia and platysma.

Operation. Incision one inch, just on the jaw, along the anterior border of the masseter muscle; vein lies posteriorly. Pass the thread from behind forward. *Guides.* Anterior edge of masseter muscle, and groove in the submaxillary bone.

Occipital arises from the external carotid opposite the facial, and passes backwards under the posterior belly of the digastric, the stylo-hyoid, and the lower part of the parotid gland, across the internal carotid artery, internal jugular vein, and the pneumogastric and spinal accessory nerves. The hypoglossal nerve hooks around it beneath the gland. The artery ascends the neck to the level of the transverse process of the atlas, passes through a groove on the mastoid process of the temporal bone, beneath the sterno-mastoid, splenius, digastric, and trachleo-mastoid, pierces the insertion of the splenius, and becomes superficial.

Operation. Point of election. Occipital portion. *Incision* from the apex of the mastoid process backward and very little upward for two inches. Divide skin, superficial fascia, deep fascia, and outer border of the sterno-mastoid, the splenius, the complexus. *Guides.* Transverse process of the atlas, and the mastoid process; the artery is found between the two, and can be traced outward to a more superficial position. Isolate from the occipital vein, and ligate.

Temporal. A terminal of the external carotid. It lies in the space between the condyle of jaw and external auditory meatus.

Line. Directly upward, between the condyle of jaw and the cartilage of the ear.

Chief relations. Anterior. Branches of facial and auriculo-temporal nerves. *Posterior.* Vein, and facial and auriculo-temporal nerves. As it crosses the root of the zygoma, the artery is covered by a dense fascia derived from the parotid gland, this should not be opened.

Operation. Incision vertical, one inch long, between the cartilage of the ear and the condyle of the jaw. Skin, superficial fascia, and some fibres of the attrahens aurem are divided, artery freed, and thread passed from behind forward.

Subclavian. On the right side from the innominate. On the left side from the arch of the aorta. Three portions—

1. From its origin to inner border of scalenus anticus. This portion gives off the thyroid axis, the vertebral, and the internal mammary arteries.

2. Behind the scalenus anticus. Gives off superior intercostal artery on the right side.

3. Outer edge of scalenus anticus to lower border of first rib. Point of election is *the outer third*.

Relations of the outer third. Posterior. Scalenus medius. *Above and external.* Brachial plexus. *Anterior and below.* Subclavian vein. *Internal.* Edge of scalenus anticus. *Structures lying in front.* Skin, superficial fascia, platysma, deep fascia, a plexus of veins formed by the external jugular, suprascapular, and transversalis colli; clavicle and subclavius muscle; suprascapular artery.

Operation. Position of patient, recumbent, shoulder supported on pillows, head back, face toward sound side, arm of the affected side depressed as much as possible. *Superficial guide*, most prominent part of clavicle. *Deep guides*, brachial plexus above and behind, outer edge of scalenus anticus muscle, and tubercle of first rib internal. *Incision.* The skin is drawn down from the neck over the clavicle, and a three-inch incision made upon the bone, from the external border of the sterno-mastoid muscle outwards. On releasing the skin this wound lies somewhat above the clavicle. Secure or push aside the external jugular vein, open the deep fascia, feel for the tubercle of the first rib and the outer border of the anterior scalene muscle; free the artery by blunt dissection, and pass the thread from below.

Collateral circulation. Suprascapular artery and posterior scapular, branch of the transversalis colli with the subscapular and circumflex. Internal mammary, superior intercostal, and aortic intercostals, with the long and short thoracics.

First part of subclavian artery. Right side. In front. Skin, superficial fascia, platysma, and deep fascia. Three muscles, sterno-mastoid, sterno-hyoid, sterno-thyroid. Three veins, internal jugular, vertebral, anterior jugular. Three nerves, vagus, cardiac filaments of sympathetic, phrenic. *Behind.* Longus colli, and three nerves, sympathetic cardiac branches of vagus and recurrent laryngeal. *Below.* Pleura and recurrent laryngeal.

Left side. Longer, more deeply placed, ascends almost vertically to neck. *In front.* Pleura, lung, internal jugular and innominate veins, the same muscles and nerves as on the right side. *Behind.* Œsophagus, thoracic duct, and as on right side, except the recurrent laryngeal. *Inner side.* Œsophagus, trachea, thoracic duct. *Outer side.* Pleura and lung.

Second part of the subclavian. Rests between the anterior and middle scalene muscles, with brachial plexus *above;* phrenic nerve, transversalis colli and supra-scapular arteries *in front;* and pleura *below.*

Internal mammary. Arises from the first portion of the subclavian and passes down behind costal cartilages to sixth interspace. *Line of incision* is vertical, two and one-quarter inches long, beginning at lower border of clavicle one-quarter of an inch external to margin of sternum; or the incision may be transverse. The point of election is in the first three intercostal spaces.

Chief relations. Anterior. Costal cartilages and internal intercostal muscles. *Posterior.* Pleura. As it is about to enter the chest it is crossed by the phrenic nerve.

Axillary. Continuation of the subclavian. *Extends* from the lower border of the first rib to the lower border of the insertion of the teres major.

Course. With abducted arm, from the middle of the clavicle to the inner border of the coraco-brachialis muscle. Three portions—

1. Lower border of first rib to upper border of pectoralis minor. *Branches.* Superior thoracic, acromio-thoracic; the latter runs along the upper border of the pectoralis minor.

2. Behind pectoralis minor. *Branches.* Long thoracic, at the lower border of the pectoralis minor, alar thoracic.

3. From lower border of pectoralis minor to insertion of latissimus dorsi and teres major. Branches, subscapular running in the posterior axillary fold, posterior circumflex, anterior circumflex.

Points of election. First and third portions, particularly the last.

Operation. First part. Patient supine, arm carried from the side. Incision three inches, commencing one-half inch from the sterno-clavicular articulation, extending outward along the line between the sternal and clavicular portions of pectoralis major. Work upward and backward between the two portions of the pectoral muscle till a dense fascia, the costo-coracoid, is reached; depress the shoulder and tear the fascia with the director, when the axillary vein is found; behind it is the artery, and still deeper the brachial plexus. Pass the ligature from below. *Guides.* The brachial plexus *behind and above.* Subclavian vein, *below and in front. Inner border of pectoralis minor, externally.*

Third portion. Arm abducted and supinated. Incision three inches long, in the hollow of the armpit, along a line passing from the junction of the anterior and middle third of the axilla to the middle of the bend of the elbow. Divide skin, superficial and deep fascias; relax by bending the elbow, displace the median nerve to the outer side, the axillary vein with the ulnar and internal cutaneous nerves to the inner side. Open the sheath, and pass the thread from the inner side.

Relations. In front. Skin and fascia only at lower part of its course. At the upper part, pectoralis major, internal cutaneous nerve, inner head of median. *Behind.* Subscapularis, tendon of latissimus dorsi and teres major, musculo-spiral and circumflex nerves. *Outer side.* Coraco-brachialis, median nerves, musculo-cutaneous nerve. *Inner side.* Ulnar nerve, nerve of Wrisburg, axillary vein. *Guides.* Superficial, the coraco-brachialis. Deep, the branches of the brachial plexus.

Collateral circulation. Ligation of first part. Acromio-thoracic and superior thoracic with subscapular and circumflex. Long thoracic with intercostals and internal mammary.

Ligation of third part. Posterior circumflex and subscapular with superior profunda; anastomoses through muscular branches and through the bone.

Brachial. Continuation of the axillary, from the lower border of the teres major, along the inner and anterior aspect of arm to one-half inch below the bend of the elbow. Passes along the inner border of biceps and coraco-brachialis, which are its *muscles of reference*, or guides.

Chief relations. Anterior. Skin and fascia; at middle third median nerve; at lower third, bicipital fascia with median basilic vein resting on it. *Posterior.* Long head of triceps, insertion of coraco-brachialis, brachialis anticus, musculo-spiral nerve, superior profund artery. *Inner side.* Internal cutaneous and ulnar nerves, median nerve (below), basilic vein. *Outer side.* Median nerve (above), coraco-brachialis and biceps. The median nerve first to the outer side, passes in front, then to the inner side. *Branches,* 1 muscular, 2 superior profund, accompanying musculo-spinal nerve, 3 inferior profund, accompanying the ulnar nerve, 4 nutrient, 5 anastomotica magna.

Operation. Arm extended and everted. Incision three inches, along the inner border of the biceps, or in the line of the artery (from the junction of the anterior and middle third of the axilla, to the middle of the bend of the elbow). Avoid the median basilic vein if it lies in the superficial fascia at the seat of operation.

At the bend of the elbow. Incision three inches. One-half inch internal to the tendon of the biceps, the lower end lying over the neck of the radius. Divide skin, superficial fascia, bicipital fascia, avoiding or tying the median basilic vein. The artery is exposed, lying upon the brachialis anticus, with the biceps tendon to its outer, the pronator radii teres muscle to its inner side.

Collateral circulation. Circumflex and subscapular with superior profund; profund with radial ulnar and interosseous recurrents.

Radial. A terminal of the brachial, passes from one-half inch below bend of elbow, along radial side of forearm to wrist, winds backwards around outer side of

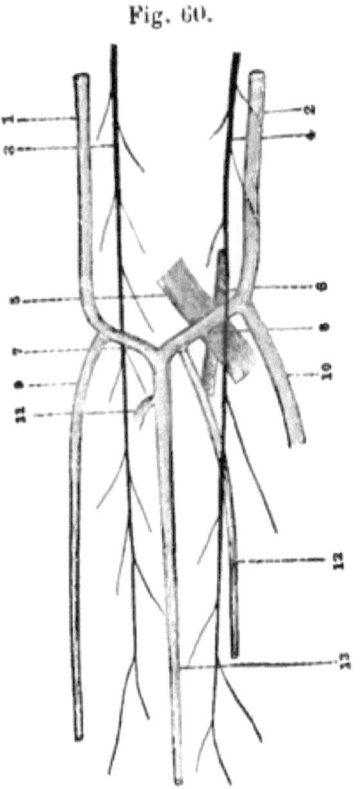

Fig. 60.

Relation of brachial artery to bicipital fascia, internal cutaneous nerve, and median basilic vein at the bend of the elbow.

carpus beneath extensors of thumb, and enters palm of hand beneath the two heads of the first dorsal interosseous muscle. *Line.* From middle of bend of elbow to a point midway between tendon of flexor carpi radialis, and styloid process of radius. *Guide.* Inner border of supinator longus.

Chief relations. Upper third. External, supinator longus muscle ; *internal,* pronator radii teres. *Lower two-thirds. External,* supinator longus ; *internal,* flexor carpi radialis. In the middle third the radial nerve is to the radial side of the artery.

Operation. Division of skin and fascial only ; the artery is superficially placed in the muscular interspace.

Ulnar. A terminal of the brachial. Commences one-half inch below middle of bend of elbow, crosses obliquely to ulnar side of arm, and continues along its ulnar border to the wrist.

Line. From a point at junction of upper and middle thirds of forearm, and three-fourths of an inch external to ulnar border, to the radial border of pisiform bone.

Chief relations. Below, flexor profundus digitorum ; *external,* flexor sublimis digitorum ; *internal,* flexor carpi ulnaris and ulnar nerve. In the upper third of its course it lies beneath the superficial set of flexor muscles. In the lower two-thirds, in its muscular interspace beneath the superficial and deep fascia only.

Operation. Pass the needle from within outwards. *Guide*—flexor carpi ulnaris.

Palmar arches. *Superficial.* Direct continuation of the ulnar artery, convex downwards, completed by the superficialis volæ of the radial, or the radialis indicis. Beneath it lie the digital arteries, nerves, and tendons of the flexor sublimis digitorum.

Deep. The direct continuation of the radial, completed by the profunda branch of the ulnar ; it rests upon the palmar interossei, and metacarpal bones near their carpal ends. It lies beneath the arteries, nerves, and tendons of both superficial and deep flexors.

Position of the arches. The superficial lies in a line drawn directly across the palm of the hand, from the angle of junction of skin covering the inner border of the thumb and the outer border of the metacarpal bone of the index-finger. The deep arch lies a finger's breadth nearer the wrist.

External iliac. A branch of the common iliac. Its course is represented by the lower two-thirds of a line drawn from three-fourths of an inch below and to the left side of the umbilicus, to a point midway between the anterior superior spinous process of the ilium and the symphysis pubis. Just above Poupart's ligament it gives off the deep epigastric, and the deep circumflex iliac.

Chief relations. Anterior. Peritoneum, spermatic vessels, vas deferens, genital branch of genito-crural nerve, circumflex iliac vein. *Posterior.* Psoas magnus and, on the right side, the external iliac vein. *External.* Psoas magnus. *Internal.* External iliac vein and vas deferens.

Operation. Patient recumbent, shoulders raised, knees and thighs flexed. *Incision.* From one inch above anterior superior spinous process ilium, to external abdominal ring, parallel to Poupart's ligament. Pass the needle from within outwards, and avoid including the genital branch of the genito-crural nerve.

Collateral circulation. Gluteal and obturator with external circumflex. Sciatic with superior perforating and circumflex branches of profunda. The deep circumflex iliac with the ilio-lumbar, the lower intercostals, and the lumbar branches of the aorta. Internal pudic with the external pudic and internal circumflex. Mammary, inferior intercostals, and obturator with deep epigastric.

Femoral. The direct continuation of the external iliac, and extends from the middle of Poupart's ligament to the opening in the adductor magnus. Its upper part is a little internal to the head of the femur; its lower part lies to the inner side of the shaft of the bone.

In Scarpa's triangle it is superficial. Below it is more deeply seated, and is in Hunter's canal.

Line. From middle of Poupart's ligament to inner side of internal condyle.

LIGATION OF ARTERIES.

Branches. Superficial epigastric, superficial circumflex iliac, external pudic, profunda, femoris, anastomotica magna.

Point of election. Apex of Scarpa's triangle.

Relations. Behind. Psoas, pectineus, femoral vein, adductor longus, adductor magnus. *Inner side.* Femoral vein, adductor longus, sartorius. *Outer side.* Psoas, vastus internus, femoral vein, internal cutaneous and long saphenous nerves. *In front.* Skin, superficial and deep fascia, internal cutaneous and long saphenous nerves, sartorius. The vein lies first to the inner side of the artery, at the apex of Scarpa's triangle behind, in Hunter's canal to the outer side.

Operation. Point of election. Thigh flexed and rotated outward, knee bent. Incision four inches in the course of the vessel, its centre at

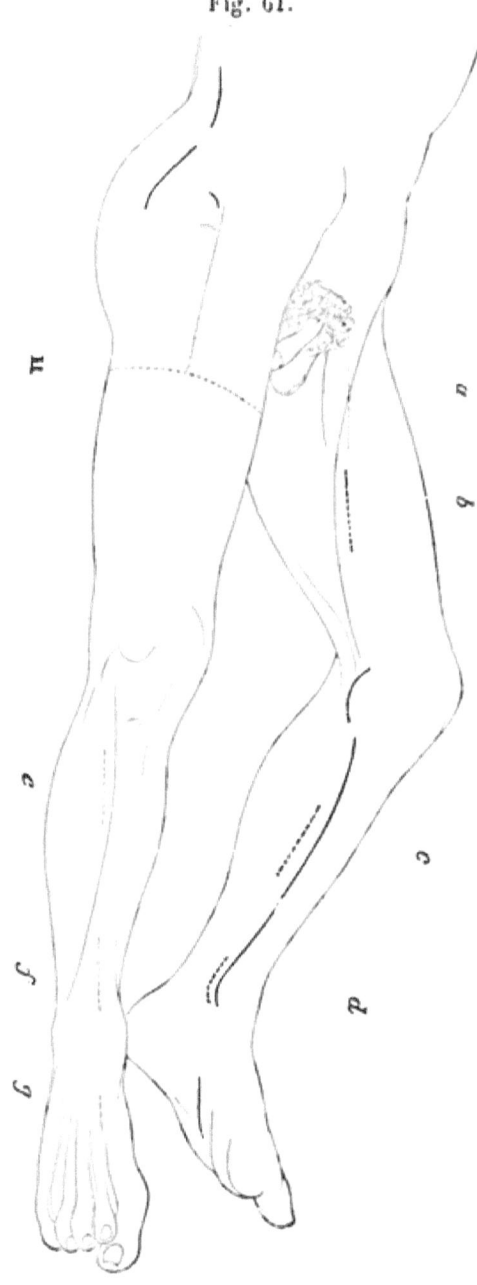

Fig. 61.

Lines of incision for ligation of femoral, tibial, and dorsalis pedis arteries.

the apex of Scarpa's triangle. On dividing the deep fascia, draw the sartorius outwards. The sheath of the vessel is cleared, and the thread passed *from* the vein.

Hunter's canal. Incision four inches exactly in the middle third of the thigh, and somewhat internal to the line of the artery. Draw the sartorius inwards, open Hunter's canal from above, avoiding the long saphenous nerve, free the artery, and pass the thread from without inwards.

Scarpa's triangle is a space situated at the upper third of the anterior surface of the thigh. *Base*, Poupart's ligament. *Outer boundary*, inner border of sartorius. *Inner boundary*, adductor longus. *Roof*, skin, superficial, deep and cribriform fascia. *Floor*, iliacus, psoas, pectineus, adductor longus, and adductor brevis. *Apex*, crossing of sartorius and adductor longus. *Length*, from base to apex, four inches.

Hunter's canal. A triangular, aponeurotic canal, corresponding to the middle third of the thigh. *Anterior*, sartorius. *External*, vastus internus. *Internal*, adductor magnus. This canal incloses the femoral artery, vein, and long saphenous nerve.

Collateral circulation. Common femoral. Gluteal, circumflex iliac and ilio-lumbar with the external circumflex. Obturator and sciatic with internal circumflex. *At apex of Scarpa's triangle.* Comes nervi ischiadici with arteries of the ham. Perforating branches of profunda femoris and anastomotica magna with articular arteries of popliteal, and recurrent of the anterior tibial.

Popliteal. A continuation of the femoral, from the opening in the adductor magnus. It passes obliquely downwards and outwards behind the knee-joint, and ends at the lower border of the popliteus muscle. The artery, throughout its extent, lies in the popliteal space. It lies deep, and is crossed by the internal popliteal nerve and the popliteal vein. The nerve lies superficial to the vein, which, in turn, is superficial to the artery.

Line. Middle of ham; the vessel runs along the external border of the semi-membranous tendon.

Relations. Upper third, from outer side, 1. Nerve. 2. Vein. 3. Artery. *Lower third* from outer side, 1. Artery. 2. Vein. 3. Nerve. *Branches*, 4 articulars, 2 muscular, azygos, cutaneous.

LIGATION OF ARTERIES.   275

Operation. Rarely undertaken. Patient supine, leg extended. Incision four inches, in the line of the artery. Great care must be exercised in separating the vein from the artery. In operating on the lower third, avoid the external saphenous vein.

Collateral circulation. Articulars with anastomotica magna and external circumflex. Superior muscular branches with terminals of profund.

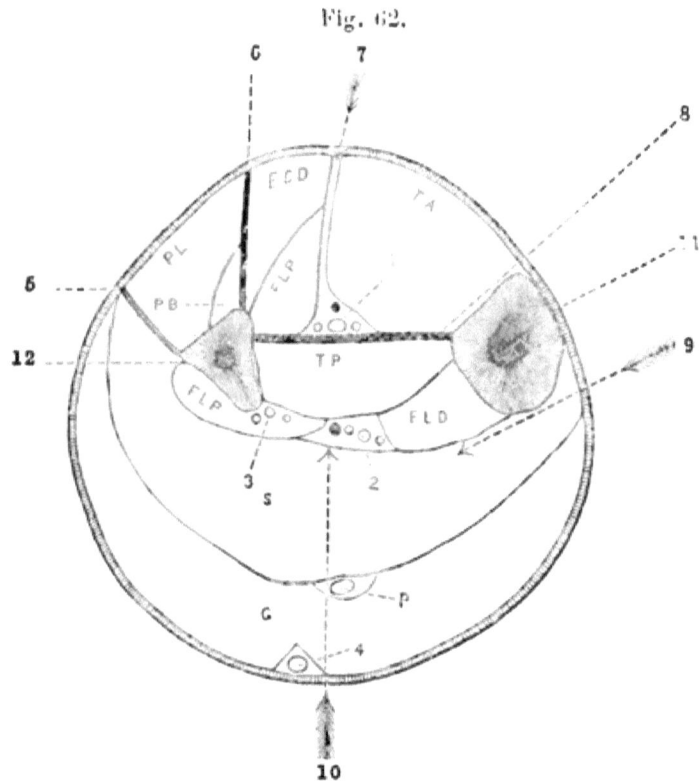

Fig. 62.

The arrow marks the tendinous arch between the flexor longus pollicis and flexor longus digitorum, beneath which the posterior tibial artery lies.

Posterior tibial. From the popliteal, at the lower border of the popliteus muscle (corresponding to the level of the lower part of the tubercle of the tibia), to a point a finger's breadth behind the external malleolus. The vessel is covered by skin and fascia, gastrocnemius, soleus, plantaris, and a tendinous arch extending between the flexor longus digitorum and the flexor longus pollicis. The posterior tibial nerve crosses the artery in its upper portion, from the inner to the outer side. The artery rests upon the tibialis posticus, the flexor longus digitorum, and the lower end of the tibia.

Line of incision. *Upper third,* along inner border of tibia. *Middle third,* one-half inch from inner border of tibia. *Lower third* (ankle), midway between internal malleolus and tendo Achillis. Pass the ligature from the nerve. Incision in upper and middle third four inches. The artery in its upper third lies very deep, and is secured by separating the soleus from the tibia working outwards in the muscular interspace between the soleus and the flexor longus digitorum.

Behind malleolus. Incision two inches long, a finger's breadth behind the internal malleolus, convex backward. Artery lies beneath the deep fascia. *Relations. Anterior.* Tendon of flexor longus digitorum. *Posterior.* Nerve and tendon of flexor longus pollicis. *Branches.* Nutrient, peroneal, muscular, communicating calcanean.

Anterior tibial. Commences at the lower border of the popliteus muscle, passes forwards between the two heads of the tibialis posticus, through an opening above the interosseous membrane to the deep part of the front of the leg, descends on the anterior surface of the interosseous membrane (upper two-thirds), and tibia (lower one-third), to the middle of the bend of the ankle joint, where it is more superficial and becomes the dorsalis pedis.

Line. From a point midway between the tubercle of tibia and head of fibula to the centre of the intermalleolar space.

The ligature is passed from the outer side.

Relations. Upper third. Between the tibialis anticus and extensor longus digitorum. Nerve to outer side. *Middle third.* Between tibialis anticus and extensor proprius pollicis. Nerve in front or to inner side. *Lower third.* Between extensor proprius pollicis and extensor longus digitorum, or frequently as in middle third. Nerve to outer side.

Operation. Upper third. Patient supine. Knee flexed, sole of foot resting on table. Incision three inches. After opening deep fascia search with handle of knife for interspace between tibialis anticus and extensor communis digitorum; artery found between them resting on interosseous membrane. Nerve to outer side. Pass thread from without. The *interspace* may be defined by extending the toes and the foot in turn, thus putting each muscle upon the stretch. Middle and lower third, as for upper third, except for the changed relations. *Branches.* Anterior tibial recurrent, muscular, internal malleolar, external malleolar.

Dorsalis pedis. The continuation of the anterior tibial. Extends from the centre of the instep beneath the annual ligament, to the base of the metatarsal bone of the great toe, where it divides into the communicating and dorsalis hallucis. Its course is from the centre of the instep, to the space between the first two toes.

It is covered simply by skin and fascia, and crossed near its point of bifurcation by the innermost tendon of the extensor brevis digitorum, which serves as a guide in its ligation.

The ligature is passed from without inwards. The artery is found between the tendon of the extensor proprius pollicis and the inner tendon of the extensor brevis digitorum. Anterior tibial nerve lies to the outer side. Incision one inch long.

External plantar artery, a terminal branch of the posterior tibial. Passes from the lower part of the internal lateral ligament posterior to the internal malleolus, forward and outward, taking a slightly arched course with the convexity outward, to the base of the fourth metatarsal space. This forms its superficial part, and is covered by the fasciæ and first layers of the foot muscles. From this point it winds round the outer border of the accessorius, and passes forward and inward to the posterior part of first interosseous space, forming the *plantar* arch, and lying upon the interossei, and bases of the metatarsal bones.

EXCISION OF JOINTS.

What is the distinction between excision and resection?
Excision means the removal of the joint surfaces of bone. Resection means the removal of the shaft of a long bone.

What is arthrectomy?
The removal, by dissection, of the diseased synovial membrane of a joint, without interfering with the bone.

What conditions may require excision?
Injury. Instance, compound luxation, compound comminuted fracture.

Disease. Instance, tubercular synovitis or arthritis.

Deformity. Instance, anchylosis in bad position.

What conditions contraindicate excision?
Malignant growth. Acute disease. Extensive involvement of bone or soft parts. Extremes of age. Marked amyloid degeneration.

What precautions are observed in excising a joint?
The incision should be free, and in the long axis of the limb. Spare the bone, substituting the gouge or curette for the saw whenever practicable. Save the periosteum and the capsule of the joint, if they are healthy. Secure absolute immobility by splinting.

How do you dress an excision?
Bone drainage-tubes, iodoform, protective, bichloride gauze, bichloride cotton, plaster bandage. Where a movable joint is desired, do not apply the fixed dressing.

Shoulder-joint. Position of patient, on his back, the affected shoulder projecting beyond the side of the operating table.

Incision four inches in length from a point slightly above and to the outer side of the coracoid process, downward and somewhat outward, external to the cephalic vein. The long head of the biceps should be freed by a longitudinal cut. The humerus is rotated outwards, and the periosteum and tendon of the subscapularis separated by the elevator. The humerus is then rotated inwards, and the periosteum and muscular attachments to the greater tuberosity are separated. Finally the humerus is forced directly upward, the posterior part of the capsule is freed by the periosteal elevator (avoid the posterior circumflex artery and circumflex nerve), the bone is sawed through the surgical neck. A posterior opening is made for drainage, and the wound dressed with a pad in the axilla and the arm to the side. Motion as soon as possible.

Elbow-joint. Incision three to four inches long, slightly internal to the middle line of the olecranon and humerus, with its central point opposite the top of the olecranon. Clear the olecranon of periosteum and soft parts with the elevator (carefully guarding the ulnar nerve) and saw off; now forcibly flex the humerus and clear it in the same way, sawing from before backward, just above the trochlear surface. Finally clear the ends of the radius and ulnar, and remove their articulating extremities just below the sigmoid notch and capitellum. *Strip the bones subperiosteally.*

Wrist-joint. Two incisions. *The radial incision*, planned to avoid the artery, commences at the level of the styloid process, on the middle of the dorsal aspect of the radius, passes downward, parallel to the tendon of the extensor secundi internodii pollicis, till it reaches the line of the border of the second metacarpal bone; it is then carried longitudinally downward for half the length of the bone.

The ulnar incision. From a point two inches above the lower extremity of the ulna and just anterior to the inner edge of the bone, downward as far as the middle of the fifth metacarpal bone.

Hip-joint. *Anterior incision*, three inches long, running downward and slightly outward, from half an inch below and external to the anterior superior spinous process of the ilium.

Fig. 63.

Metacarpal saw.

280 ESSENTIALS OF SURGERY.

Posterior incision. Begin midway between anterior superior spine of ilium and top of trochanter; sweep backward and downward behind posterior margin of the trochanter for about three inches, keeping about an inch back of the edge of the bone. *Do not force the head of the bone from the wound*, but divide *in situ* by a narrow saw; remove subsequently with sequestrum forceps. Curette and gouge away all diseased portions of the acetabulum, remove diseased synovia or capsule, wash out with zinc chloride, dry with bichloride sponges, dust with iodoform. Dress antiseptically and apply a double Thomas's splint.

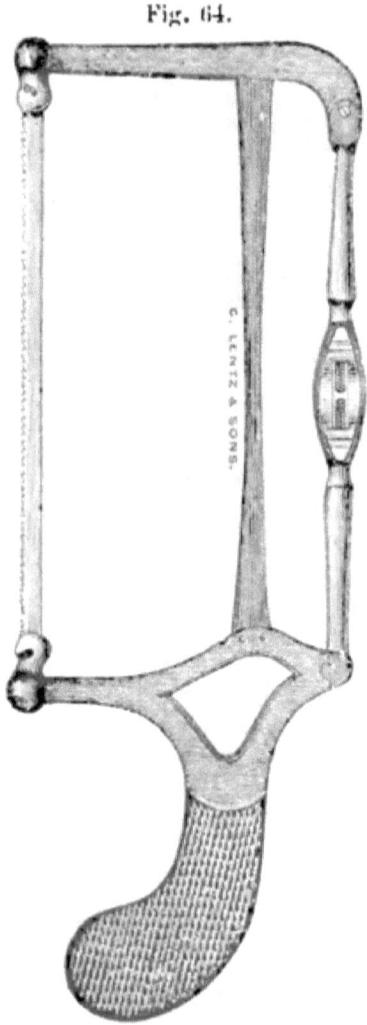

Fig. 64.

Butcher's saw.

Knee-joint. Incision from the outer and posterior border of the internal condyle, to a corresponding point on the external condyle, curving downward sufficiently to pass midway between the patella and the tuberosity of the tibia. Dissect up the anterior flap containing the patella, flex the joint, divide the lateral and crucial ligaments, clear the end of the femur with the finger, saw at right angles to its long axis near the upper margin of the cartilaginous surface. Use Butcher's saw, cutting from behind forward. Clear the end of the tibia, and remove its articulating extremity. Remove by the gouge or curette all diseased tissue. Suture the bone together with thick cat-gut or silver wire, provide for drainage, and close. Absolute fixation, plaster bandages if the wound remains aseptic.

Ankle-joint. Very rarely performed. Every effort should be made to preserve the periosteum. Two incisions are made. *The fibular* begins two-and-a-half inches above the tip of the external malleolus, passes downward along its posterior border, around its tip, and upwards along the anterior border for an inch (hook-shaped). *The tibial* forms a semicircle around and just below the internal malleolus, from the middle of which a third cut runs directly upwards over the malleolus for two inches (anchor-shaped). The periosteum is first raised from the fibula, when the bone is sawed and removed. Next, the articulating end of the tibia is removed; finally the astragalus is sawn through. If the elevator is carefully used, the tendons and their sheaths will not be damaged.

AMPUTATIONS.

Under what circumstances is amputation required?

1. Avulsion of a limb. 2. Mortification. 3. Compound luxations and fractures, if seriously complicated. 4. Extensively lacerated and contused wounds. 5. Diseases of bones and joints. 6. Lesions or diseases of arteries. 7. Morbid growths. 8. Deformity.

What instruments are required in amputation?

Tourniquets, knives, saws, retractors, tenacula, artery forceps, hæmostatic forceps, bone-nippers, scissors, needles, and sutures.

Describe the methods of operating.

1. *Circular.* The skin is drawn upward and divided by a circular sweep of the knife, passing entirely around the limb, and

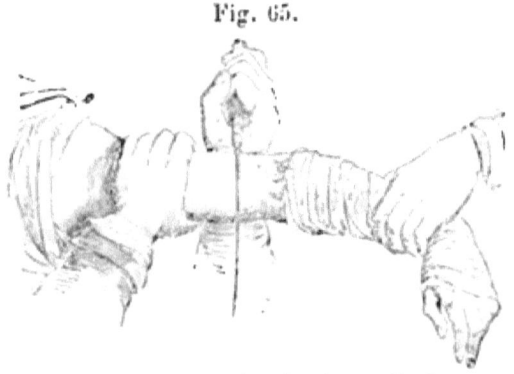

Fig. 65.

Amputation by the circular method.

dividing everything down to the muscles; this skin cuff is further dissected up till its length is a little greater than half the diameter of the limb; it is then retracted, the muscles are separated down to the bone by a second circular incision, and the latter is sawed through.

2. *Flap.* There may be one or two flaps; these may be anterior, posterior, lateral, square or oval; they may be cut by transfixion, or from without, and may include all the soft parts (mus-

Describe the methods of shaping the flap.

Modified circular. Two short, curved, skin-flaps are cut, and the notched skin cuff is dissected up as in the circular method.

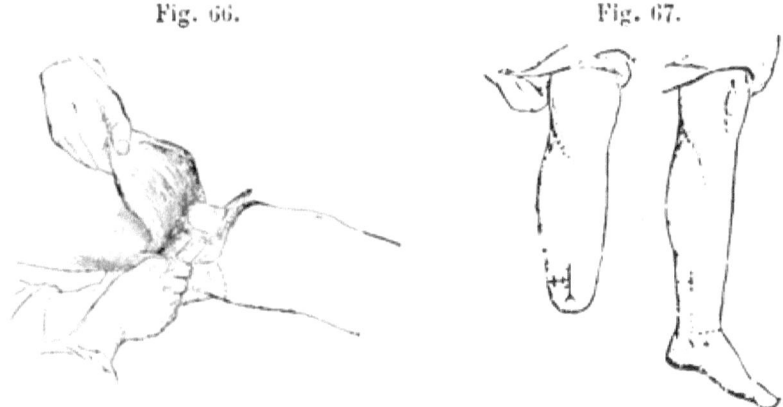

Fig. 66. Formation of flaps by transfixion.   Fig. 67. Teale's amputation.

Oval and elliptical. The oval method is practically a circular incision, with the cuff slit at one side, and its angles rounded off.

In the elliptical method the incision forms a perfect ellipse; the flap is folded upon itself and sutured, making a curved cicatrix.

Teale's method. Rectangular flaps, each equal in breadth; one has a length of half the circumference of the limb, the other (containing the bloodvessels) is only quarter as long.

How are amputations classified in regard to the time of operating?

Primary, before the occurrence of inflammatory fever. *Intermediate,* during acute inflammatory fever. *Secondary,* after suppuration has been established.

What period is most favorable for amputation?

Before the occurrence of inflammatory fever. If the time for primary amputation has passed, wait for the *secondary* period.

What sequelæ may occur after amputation?
Hemorrhage, muscular spasm, pain, inflammation, osteomyelitis, protrusion of bone.

Amputations of the Foot.

Lisfranc's amputation. Tarso-metatarsal disarticulation; between the metatarsal bones and the three cuneiforms and cuboid.

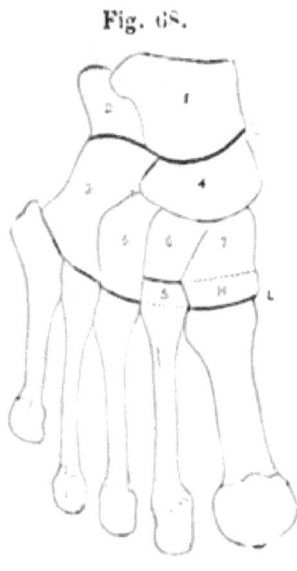

Fig. 68.

L. Lisfranc's operation. H. The extremity of the internal cuneiform removed by Hey's operation. C. Chopart's operation.

Incision. From the base of the first to the base of the fifth metatarsal bone across the dorsum of the foot, with a marked convex curve downward. Forcibly extend and disarticulate, bearing in mind the backward projection of the second metatarsal bone. Cut a long plantar flap.

Arteries. Dorsalis pedis and plantar arches.

Hey's amputation. The same as Lisfranc's, except that the projecting internal cuneiform bone is sawed through.

Chopart's amputation. Intertarsal disarticulation, between the astragalo-scaphoid, and calcaneo-cuboid joint.

Incision. From a point midway between the tuberosity of the fifth metatarsal bone and the external malleolus, a curved dorsal incision is made to a point one-half inch behind the tubercle of the scaphoid. Extend the foot, disarticulate, and cut a long plantar flap.

Pirogoff's amputation. Through the ankle-joint and os calcis.

Incision. from the tip of the external malleolus, across the under surface of the heel, to a point half an inch below and behind the internal malleolus. Incline this cut well forward. Forcibly extend the foot and unite the ends of the first incision by a deep cut passing directly across the dorsum. Open the joint, draw the foot forward, place a narrow saw behind the astragalus and saw the os calcis through in the line of the first skin incision. Saw off the ends of the tibia and fibula, bring the heel flap up till the sawn bone surfaces are in contact, unite them with heavy catgut, and suture the wound.

Syme's amputation. Through the ankle-joint.
Incision. Inclining *backward* from tip of external malleolus, beneath the heel, to a point half an inch below and behind the internal malleolus. Dissect the flap from the os calcis *cutting towards the bone.* Unite the ends of the first incision by a transverse cut across the front of the ankle-joint, disarticulate, saw off the articular extremities of the tibia and fibula, and bring the flaps together.

Amputations of the Leg.

Lower third of the leg. By the circular, modified circular, bilateral tegumentary flap, Teale's method. The fibula should be divided first. *Arteries.* Anterior and posterior tibial, peroneal, and muscular.

Middle and upper third of the leg. By a long anterior tegumentary flap half the circumference of the limb in breadth and a little more in length. By short antero-posterior flaps. By lateral musculo-tegumentary flaps (Sédillot's). The projecting sharp edge of the tibia should be covered with a flap of periosteum to prevent perforation of the anterior flap.

Lateral double flap method (Sédillot's). A long external flap is formed by transfixion, and united to the short internal flap formed by the calf muscles.

Lateral tegumentary flaps may be formed cutting from without inward.

Point of election in leg amputation. Two inches below the tuberosity of the tibia.

Amputations at the Knee-Joint.

Where indicated by injury or disease this is one of the most successful of all leg amputations, and leaves a far more serviceable stump than amputation in the continuity of the limb.

Lateral flap operation. Commence the incision in the middle line an inch below the tubercle of the tibia, form a flap convex downward, carrying the point of the knife to the centre of the posterior surface, when it is continued directly upward to the centre of the articulation. The second incision begins at the same point as the first, and pursues the same course on the opposite side of the leg to the posterior median line. The anterior incisions should incline forward to allow sufficient material for covering the condyles. The internal flap should have additional fulness. The patella and semilunar cartilages are allowed to remain.

Long anterior flap. Incision from the lower extremity of the inner condyle downward for three inches, then directly across the tibia and upward to the external condyle. Disarticulate and cut a short posterior flap.

Amputation through the femoral condyles (Carden's). Incision, from the upper border of the inner, to the upper border of the external condyle, carried downward and across the front of the leg just below the insertion of the ligamentum patellæ. Short posterior flap by transfixion. Condyles sawed across. The patella is not left in the anterior flap.

Gritti's modification. Consists in sawing off the articular surface of the patella, turning it backward, and suturing it to the divided femur.

Amputations of the Thigh.

Antero-posterior musculo-tegumentary flaps. Anterior cut from without inwards, about four inches long, and somewhat square. Posterior flap about the same length to allow for retraction, cut by transfixion. The posterior muscles of the thigh always retract more than the anterior group.

Lateral flap. Teale's method or **modified circular** operation may also be done on the thigh.

Hip-Joint Amputation.

Hemorrhage controlled by abdominal tourniquet, digital pressure on the femoral, and Esmarch's tube applied in the form of a spica of the groin.

Long anterior and short posterior flaps. Enter the knife at a point midway between the anterior superior spinous process of the ilium and the tip of the trochanter, push it directly across the capsule of the joint, grazing the head of the bone, till it appears on the inner side of the thigh just in front of the tuber ischii; cut directly downwards for six inches, let the femoral artery be seized by the fingers of an assistant, then complete the anterior flap by cutting outward. Turn the flap up, clear the capsule, forcibly extend the femur, and, placing the knife behind the trochanter, form a somewhat shorter posterior flap. First secure the gluteal and sciatic vessels, then the femoral artery and vein. The flaps may be cut from without inwards, securing the vessels as cut.

Vertical and circular method. A vertical incision is made, from a little above the tip of the trochanter for five inches in the long axis of the femur. Through the incision disarticulation is effected, and by means of the elevator and knife the soft parts are separated from the bone. At the lower extremity of the vertical incision, skin, fascia, and muscles are divided by a circular sweep of the knife around the thigh, and the entire femur, together with the soft parts below the circular cut, is removed. This operation is tedious, but far more safe than the double flap method.

Amputation of the Hand.

Phalanges. The palmar flexure is the guide to the joint surface. Flex the joint, open it by a slightly convex dorsal incision a little below its most prominent part, and cut a long palmar flap. The digital arteries can usually be secured by the skin suture. The proximal phalanx of the middle and ring fingers should not be saved.

Metacarpo-phalangeal. *Oval method (en raquette).* The point of the knife is entered in the mid dorsal line, a little above the knuckle, carried first downward, then around the side of the finger, across its web and palmar surface, and back to the point of starting.

Any of the bones of the hand may be amputated through their continuity by either the double flap, or the oval method.

Wrist-joint. Incision, convex downward, from styloid process of radius to corresponding process of ulna. Dissect up the flap, divide tendons, disarticulate, and cut a palmar flap from within, guarding against the knife catching on the pisiform bone.

Amputations of the Arm and Forearm.

Forearm. Modified circular, or antero-posterior flaps. Teale's method.

Arteries. Anterior and posterior interosseous, radial and ulnar.

Elbow-joint. The line of articulation is oblique, from without inward and downward, hence there will not be enough flap to cover the internal condyle if the knife is carried directly across the arm.

Long anterior and short posterior flap. Flex and supinate the forearm, raise the soft parts from the bone, enter the knife an inch below the internal condyle, and push it across the limb close to the ulna, till it appears an inch and a half below the external condyle. Make a three-inch flap, bringing the knife out sharply at the finish. Draw the skin well up and unite the two extremities of the incision by a semilunar dorsal cut. Disarticulate, either dividing the triceps, or sawing off the olecranon.

Circular method. The incision is made three to four inches below the joint.

Arm. *Circular. Flap.* Any of the methods.

Shoulder-Joint.

Oval method. (Larrey's.) Forming lateral musculo-tegumentary flaps. Enter the point of the knife to the bone just below the acromion process, and make an incision downward in the long axis of the arm for about two inches. From the end of the incision two curved incisions are carried to the anterior and posterior axillary folds, respectively. These flaps are dissected up, and disarticulation is effected by rotating the humerus outward, and dividing first the sub-scapularis, then the long head of the biceps and capsular ligament, then rotating the humerus inward and dividing the insertions of the supra- and infra-spinator and teres minor muscles. The knife is now placed behind the bone, and the two curved incisions are joined by a transverse cut, severing the axillary artery, which is controlled by the thumb of an assistant before it is divided. Hemorrhage is checked by pressure on the subclavian, Esmarch's tube, and seizure of the artery in the flap before it is cut. *Arteries.* Anterior and posterior circumflex, supra-scapular, brachial.

Single flap method. (Dupuytren's.) A long external flap is cut from the deltoid muscle, either by transfixing, or from without in.

19

BANDAGING.
The Roller Bandage.

Describe the roller bandage.

A strip of unbleached muslin, from half an inch to three inches in width, and from three to twelve yards in length. It may be made of calico, linen, or gauze. It is tightly rolled in the form of a cylinder; the rolling may be from each end, forming the double-headed bandage.

Name the parts of a roller bandage.

The initial and terminal extremities, the upper and lower borders, the internal and external surfaces, and the body of the roller.

How do you apply a roller bandage?

Fix. The body of the roller being held in the right hand, the external surface of the initial extremity is applied to the surface,

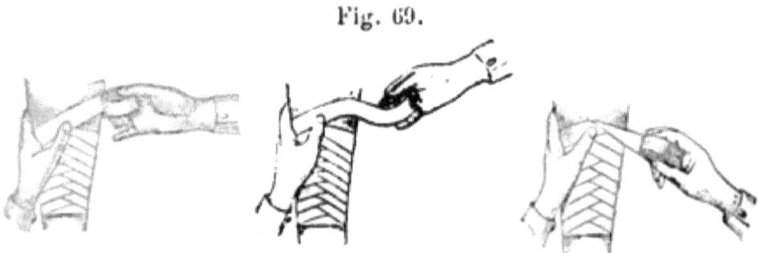

Fig. 69.

Method of applying the spiral reversed bandage.

fixed by the thumb of the left hand till it is caught by the bandage carried around the limb, when it is further held in place by a *repeated circular* turn. The following turns can be made to *overlap* this circular, covering in from a half to three-fourths of its surface. If the part is conical, the overlapping turns may be made to lie smoothly by *the reverse*.

The *circular turns* are those which pass around the part, one passing directly over the other.

The *spiral turns* are those which pass up the limb, each one overlapping the other.

BANDAGING.

The *oblique turns* are those in which the bandage passes up the limb without overlapping, leaving space between each turn.

Recurrent turns are those in which the bandage is caught, passed to and fro, across the end of a stump for instance, and the loops held at the sides by circular turns.

Spiral and figure-of-eight turns are those in which the bandage forms by oblique turns two loops in the form of an eight. By overlapping, the crossings of these loops form a series of angles or spicas.

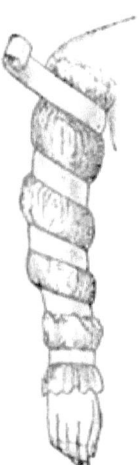

Fig. 70.

Oblique bandage.

Describe the reverse.

Consists in folding the bandage over, so that the surface in contact with the skin is changed with each reversed turn. This is accomplished by relaxing all tension on the roller, carrying the right hand, holding the body of the roller, from supination to pronation, passing the body of the roller to the left hand beneath the limb, and making firm traction.

For what purposes is the roller applied?

The general indications for all roller bandages are to retain splints and dressings, and to make pressure.

Spiral of one finger. Length, one-and-a-half yards; width, three-fourths of an inch. Fix by a circular turn at the wrist once repeated. Carry the bandage down over the dorsum of the hand, and by an oblique turn to the extremity of the finger, which is then covered in by spiral or reversed turns as required. Complete the bandage by carrying it up to the wrist, over the back of the hand, and making one circular turn.

Spiral of four fingers (gauntlet). Length, five yards; breadth, one inch. Cover in each finger precisely as above, beginning with the little finger of the left hand, the index-finger of the right. As each finger is finished, the bandage is carried to the wrist, around, and then down to the next finger. The thumb may be included in this bandage if necessary.

Spica of the thumb. Length, three yards; width, three-quarters of an inch. May be ascending or descending. *Ascending.*

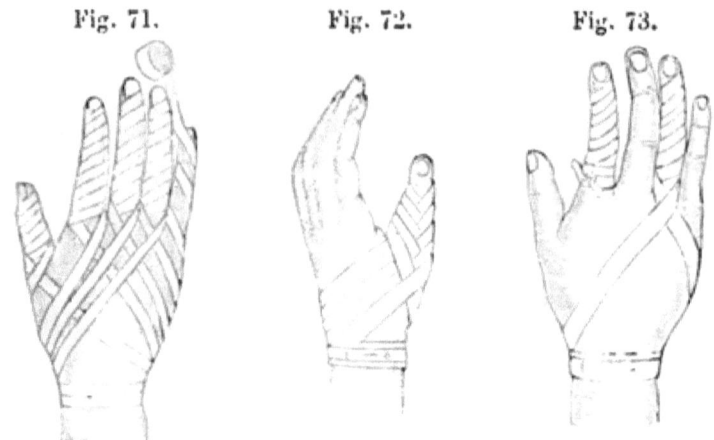

Fig. 71. Gauntlet, also taking in the thumb. Fig. 72. Spica of thumb. Fig. 73. Spiral of one finger.

Fix at the wrist. Pass to the metacarpo-phalangeal articulation, and make a circular. Pass to the wrist again, and alternate the wrist and thumb turns so that the line of crossing is over the dorsum of the thumb. Overlap two-thirds from below upward. The descending spica has the same turns, but overlaps from above downward.

Demi-gauntlet. Length, three yards; breadth, one inch. Fix at the wrist, pass obliquely across the back of the hand to the index-finger of the right hand the little finger of the left; pass around the finger, and obliquely back to the wrist. Make a circular turn, then take in the next finger in a similar way till each one is encircled by a loop.

Spiral reversed of upper extremity. Length, twelve yards; width, one and one-half inches. Apply with hand in pronation. Fix at the wrist. Carry across the back of the hand and make a circular turn about the fingers at the level of the distal joint of the little finger. Run up the hand with spiral reversed, or figure-of-eight turns, covering in the metacarpal bone of the thumb by means of the latter. Continue up the forearm with spiral turns till they cease to fit closely to the surface, when the reverses must be made. The elbow must be covered in by a figure-of-eight. Do not make the line of reverses (the line of pressure) over the subcutaneous portion of the ulna. Overlap two-thirds.

Spica of the shoulder. Length, ten yards; width, two-and-one-half inches. Ascending or descending. *Ascending.* Fix by a circular turn about the arm placed as high as possible. Carry the bandage, overlapping the circular turn where it passes over it, across the chest (right side) or back (left side), under the opposite axilla and back to the point of starting. It is now carried around the arm, overlapping the circular turn, and making a spica directly in the middle line of the shoulder with the beginning of the body turn. This is repeated, passing upward till the entire shoulder is covered in. *The descending spica* is applied by the same turns, but runs from above downward till it reaches the first circular turn.

Velpeau. Length, fourteen yards; width, two and one-half inches. For the proper application of this bandage the arm must be placed in the Velpeau position, the hand of the injured side resting on the sound shoulder.

Commence over the scapula of the sound side, carry the roller over the injured shoulder to the middle of the outer aspect of the upper arm, across the chest (behind the elbow) to the axilla of the sound side, thence to the point of starting. Repeat this turn to fix, then make a circular turn about the chest, taking in the elbow of the injured side. Repeat these turns, first shoulder, then body, overlapping so that the shoulder turns reach the point of the elbow when the body turns

Fig. 74.

Velpeau.

take in the wrist. This requires overlapping of about five-sixths for the vertical turns, one-third for the horizontal. Used to dress fractured clavicle or scapula.

Désault. Requires three rollers.

First roller. Length, five yards; width, two-and-one-half inches. It fixes a wedge-shaped pad, base up, in the axilla. Four spiral turns are made, encircling the thorax and pad, the roller is then carried from the pad obliquely to the sound shoulder, about which and the pad it is made to form a series of spica turns.

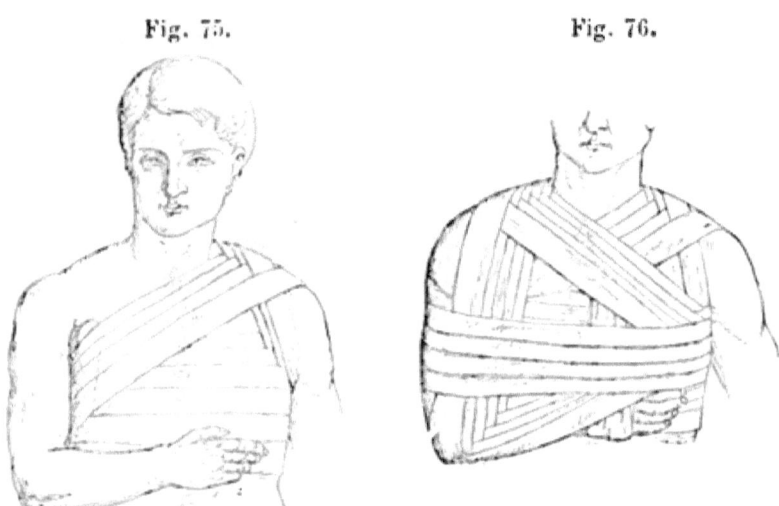

Fig. 75. Fig. 76.

Désault. First roller.   Désault. Second and third roller (the second is here applied last).

Second roller. Length, seven yards; width, two-and-one-half inches. Presses the elbow to the side, and forces the head of the humerus outward. It consists of a number of circular turns embracing the arm and chest, and running from the head of the humerus to the elbow, overlapping one-half. The upper turns are applied very lightly, as they descend the tension on each turn is increased.

Third roller. Length, seven yards; width, two-and-one-half inches. Presses the shoulder upward and backward. Begin at the axilla of the sound side, carry the roller obliquely across the chest, over the injured shoulder, down the back of the humerus, around the elbow of the injured side, across the chest again to the point of starting; then under the axilla of the sound side, obliquely across the back, over the injured shoulder, down in front of the humerus, around the elbow, across the back to the point of starting. This forms two triangles, one anterior the other posterior. Axilla, shoulder, elbow, first in front, then behind, represent the angles of the triangles. These turns may overlap two-thirds, or may exactly overlie.

Spiral of chest. Length, seven yards; width, three inches. Circular around the waist, ascends to the axilla by spiral turns overlapping one-half. Keep from slipping down by making a recurrent turn across one shoulder, pinning to the circular turns, bringing the bandage back over the other shoulder, and securing it to the circular turns in front.

Anterior figure-of-eight of chest. Length, seven yards; width, two-and-one-half inches. Fix by a circular about the right arm, then carry the roller over the shoulder, across the chest, around the left shoulder, across the chest again, around the right shoulder, across the chest, and so continue till the required number of turns have been applied. Over the sternum the spicas may run up, overlapping three-fourths.

Posterior figure-of-eight of chest. Length, seven yards; width, two-and-one-half inches. Fix the roller upon the upper part of the left arm, carry it over the left shoulder, obliquely across the back to the right axilla, around the right shoulder, obliquely across the back to the left axilla, and so continue till the necessary number of turns are applied.

Spica of breast. May be single or double.

Single. Length, ten yards; width, two-and-one-half inches. Starting from the scapula of the affected side, carry the roller over the shoulder of the sound side, just beneath the affected breast, and around the chest to the point of starting; repeat this turn, then make a circular around the chest, taking in the lower border of the mammary gland and making a spica or cross with the oblique turn. Alternate these circular and oblique turns, and continue them, overlapping two-thirds, till the gland is covered in. The spicas or crosses should all be in the same line.

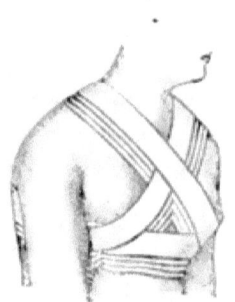

Fig. 77.

Spica of breast (double).

Double. Length, fourteen yards (two bandages); width, two-and-one-half inches. This is made up of two oblique turns to each circular. Start from the left scapula and make a repeated oblique turn, passing over the right shoulder and under the left breast as before; then carry the roller around the chest as though to make a circular turn, till it passes beneath the right breast, when it is carried obliquely upward over the left shoulder (passing above and to the inner side of the left breast); across the back, and a circular is made, just taking in the lower borders of the glands and making spicas with the two obliques.

Spica of the Foot. Length, five yards; width, two-and-a-half inches. Begin by a circular turn about the ankle; pass over the dorsum of the foot to the metacarpo-phalangeal articulation; make a circular and a spiral turn, overlapping three-fourths, then carry the roller over the dorsum of the foot to the back of the heel, around the heel, so that the lower border of the bandage extends as low as the level of the sole, then back to the dorsum of the foot, crossing the beginning of the heel turn exactly in the middle line as it overlaps the spiral turn: this forms the first spica. Again pass around the sole of the foot, across the dorsum of the foot overlapping three-quarters, around the heel, and back across the foot, making the second spica. So continue till the foot is covered in. Each turn of the bandage, after the spica is begun, must be parallel to its predecessors throughout its whole extent, and must overlap to the same degree.

Fig. 78.

Spica of the foot.

Spiral reversed of the foot covering in the heel. Length, four yards; width, two-and-a-half inches. Fix by a circular turn about the ankle, pass over the dorsum of the foot to the metacarpo-phalangeal articulation; make a circular at that point, and pass up the foot by two or three reversed turns, overlapping three-fourths; having reached the top of the instep, carry the bandage around the *point* of the heel, up over the instep, down around the *sole* of the heel obliquely, backward, and upward, below the malleolus, and around the *back* of the heel, forward to the instep. Again pass under the sole of the heel, beneath the malleolus, around the back of the heel, and forward to the instep. The bandage may be pinned at any point, or carried up the leg.

Spiral reversed of the lower extremity. Length, twelve yards; width, two-and-a-half inches. Fix at the ankle, pass down over the dorsum of the foot, and make a circular turn about the foot at the metatarso-phalangeal joint, pass up the instep by a spiral, a spiral reversed, and two or three spica turns; then pass up the leg by spiral turns, beginning to reverse as soon as the shape of the limb requires it. Cover the knee with a figure-of-eight, and ascend the thigh by spiral reversed turns. Overlap two-thirds. Do not make the line of the reverse over the crest of the tibia.

Fig. 79.

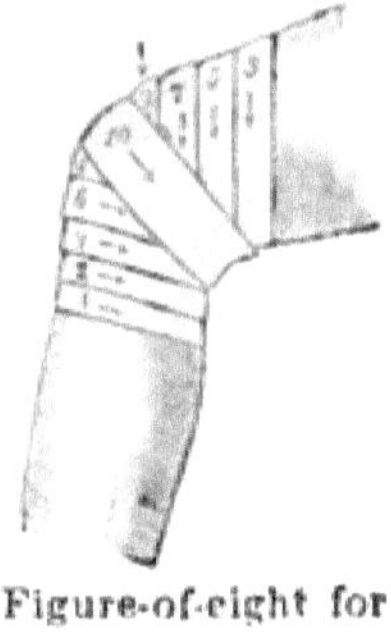

Figure-of-eight for the knee.

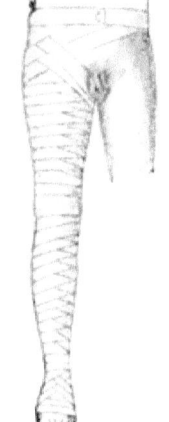

Fig. 80.

Spiral reversed of the lower extremity.

Figure-of-eight of the knee. Length, three yards; width, two-and-a-half inches. Fix by a circular three or four inches below the joint, carry the bandage upward obliquely over the popliteal space, and make a circular about the thigh, three or four inches above the joint, descend obliquely over the popliteal space, and make a circular about the leg, overlapping the first turn upward two-thirds, ascend and make a second circular about the thigh, overlapping downward two-thirds. So continue till the joint is covered.

Spica of the groin. Single or double. Ascending or descending. *Single ascending.* Length, ten yards; width, two-and-a-half inches. Fix around the upper part of the thigh (if it is the left side, the bandage must be applied throughout from right to left); carry obliquely across pubes, lower part of abdomen and crest of ilium, around the back, and down to the starting-point, passing across the front of the thigh, and forming the first spica turn, which should be within the middle of the anterior surface of the thigh; repeat these turns, overlapping two-thirds in the groin, but converging as the bandage is carried to the crest of the ilium, till they overlie in the back.

Remember that in all ascending spica bandages, the position of the crossing is determined by the lower border of the bandage; in all descending spicas, the upper border determines the position of the turns. A well-applied spica should have all the angles of crossing exactly in line.

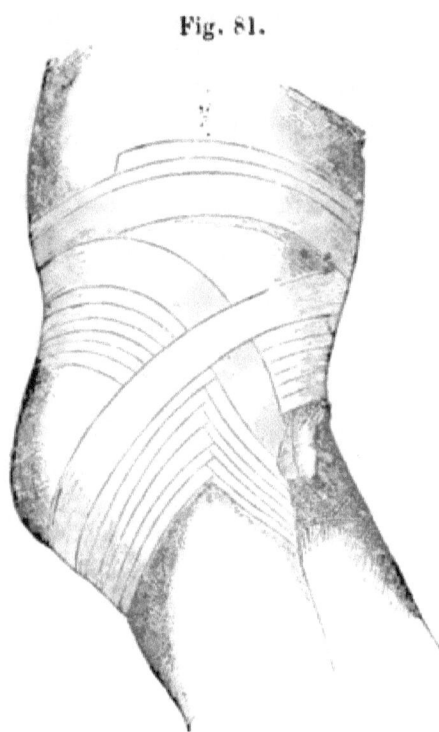

Fig. 81.

Spica of groin. Single ascending. Should be started around the thigh.

Double ascending spica. Length, fourteen yards; width, two-and-a-half inches. Fix by a circular around the waist, carry obliquely downward across the belly, pubes, and left thigh; around the left thigh, and up to the left iliac crest, forming the first spica; around the back, and obliquely down, across, and around the right thigh, forming the second spica; obliquely across the belly to the left iliac crest, forming with the first oblique abdominal turn the third spica. Repeat these turns, taking in body, left thigh, body, right thigh, and overlapping two-thirds. There are three sets of crossings: one in the middle line of the belly, and one within the middle line of each thigh.

Descending single and double spicas of groin. The turns are the same as for the ascending spicas, except that the first turns are placed at the highest point which it is desired to cover by the bandage, and the spicas are made by the upper border of the bandage.

Head Bandages.

Barton's. Length, five yards; width, two inches. Begin behind the ear (left if standing behind the patient, right if standing in front); carry the roller down under the occiput, and up to a corresponding point behind the other ear; thence directly across the vertex, down the side of the face, under the chin, up the other side of the face to the vertex, making an intersection with the former turn directly in the middle line; then to the point of starting, around under the occiput, forward along the body of the jaw, around the symphysis menti, back along the jaw on the other side, to the point of starting. Exactly repeat these turns three times. Application. Fracture of jaw.

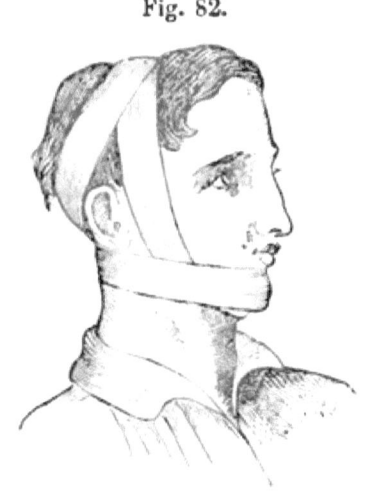

Fig. 82.

Barton's bandage.

Gibson's. Length, five yards; width, two inches. Make three vertical turns, passing under the chin, along the sides of the face in front of the ears, and over the top of the head; reverse just above the ear, and make three circular turns about the forehead and occiput; as the third turn is completed, carry the bandage beneath the occiput, under the ear, along the body of the jaw, around the symphysis menti, and take in the front of the chin and the sub-occipital region with three turns; reverse beneath the occiput, carry the roller directly forward in the middle line to the forehead, pin all intersections.

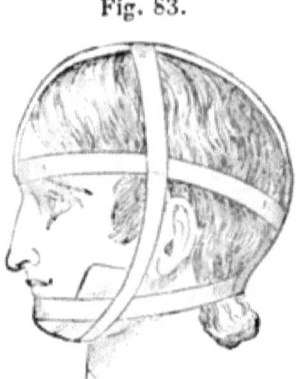

Fig. 83.

Gibson's bandage. Vertical turn should be made first.

Oblique of the jaw. Length, five yards; width, two inches. Face the patient, begin the bandage in the middle of the forehead and carry it *towards* the injured side. Fix by a circular fronto-occipital turn. Carry the roller obliquely down beneath the occiput, around the front of the neck to the angle of the injured jaw, then up the side of the face (in front of the ear), across the vertex, down the side of the head behind the ear of the sound side, under the chin, and up again on the injured side, overlapping the preceding turn forward three-quarters. The turns behind the ear of the sound side do not overlap.

Application. For fracture of the condyle of the jaw, or fractures with marked lateral deformity.

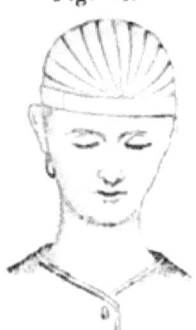

Fig. 84.

Recurrent of scalp.

Recurrent of scalp. Length, seven yards; width, two inches. Fix by a circular fronto-occipital turn, then reverse, catch the point of reverse with the finger and pass directly from occiput to brow across the top of the scalp. The bandage is held in front by an assistant and carried back again overlapping the first recurrent turn two-thirds; it is carried to and fro in this way till the scalp is entirely covered, when the loops are fixed at the sides by circular turns.

Figure-of-eight of the eye. Single and double.

Single. Length, five yards; width, two inches. Fix by a circular fronto-occipital turn, beginning in the middle of the forehead and carrying the bandage *away from* the injured eye. As the bandage passes backwards for the third turn, carry it obliquely downward across the occiput, under the ear of the affected side, obliquely upward over the ramus of the jaw and the affected eye, to the most prominent part of the parietal bone; thence to the starting-point of the oblique turn, which is to be repeated two or three times and fixed by a fronto-occipital circular. This bandage may also be applied by alternating circular and oblique turns, overlapping upward or downward and making a series of spicas.

Double. Length, seven yards; width, two inches. One eye may be covered as in the single bandage, then the other in a precisely similar manner; or the turns may alternate and overlap, forming a series of spicas over the bridge of the nose.

Occipito-facial. Simply the vertical and circular occipito-frontal turns of the Gibson bandage. Pin all intersections.

Fronto-occipito-cervical figure-of-eight. Length, three yards; width two inches. Fix by a fronto-occipital circular turn, carry obliquely downward across the occiput to the neck, around the neck, obliquely upward across the occiput, around the forehead, obliquely downward and around the neck; so continue till the bandage is completed.

Fronto-occipito-mental figure-of-eight. Length, three yards; width, two inches. Apply as the preceding bandage, except that the turn is carried around the chin instead of around the neck.

Handkerchiefs.

Describe the handkerchief bandage.

This consists of a thirty-two inch square piece of muslin, calico, or any soft strong material, forming *the square*.

The triangle is formed by bringing the two opposite angles of the square together. The parts of the triangle are, the base, the apex (the angle opposite the base), and the angles or ends.

The cravat is formed by folding the triangle once or twice from its apex towards its base.

Handkerchief bandages receive a double name, the first being the part to which the base is applied, the second the part around which the ends are carried.

The simple bandage is that made up of a *single* handkerchief; the *compound bandage* is that made up of more than one handkerchief.

Handkerchief Bandages of the Head.

Occipito-frontal triangle. Apply the base to the occiput, letting the apex fall over the forehead. Carry the two ends forward around the head and tie in front, or cross, and pin at the sides. Turn the apex up and pin to the body of the bandage.

Fronto-occipital triangle. As the preceding, except that the base is applied to the forehead, and the apex falls over the occiput.

Bi-temporal triangle. As the preceding, except that the base is applied over one temple, the apex falls over the other.

In the choice of these three bandages, the base is applied over the seat of injury, or where most pressure is desired.

Vertico-mental triangle. Apply the base to the vertex with apex back; carry the ends down under the chin, and either tie, or cross and pin. Bring the apex to one side and pin.

Auriculo-occipital triangle. This does not conform to the rule in naming. Place the base in front of the ear, apex back, carry one end under the chin, the other over the top of the head and tie or pin in front of the ear on the sound side.

Square cap. Fold the handkerchief so that a quadrilateral is formed, with one border overlapping the other three inches. Apply this quadrilateral to the scalp with the projecting border next the surface and hanging over the forehead. Bring the ends of the short fold under the chin and tie. Fold back the long border exposing the forehead, pull the ends forward till the bandage fits about the head, then carry them back and tie beneath the occiput.

Fig. 85.
Beginning of square cap of head.

Fig. 86.
Square cap of head completed.

Fronto-occipito-labialis cravat. Fold the triangle into a cravat. Place the body upon the forehead, carry the ends back, cross at the back of the neck, and bring them forward, tying or pinning over the upper or lower lip, as required by the injury. Used to approximate lip wounds, and to check bleeding from the coronary arteries.

Occipito-sternal triangle (compound). Apply a sterno-dorsal (straight around) cravat about the chest. Flex the head upon the chest and apply the base of a triangle, apex forward to the occiput, carry the two ends down to the sterno-dorsal cravat and secure. The apex of the triangles may be folded back and pinned. Used in cut throat wounds of the neck.

Parieto-axillaris triangle (compound). Apply an axillo-acromial cravat (around the shoulder). Place the base of a triangle over the parietal eminence of the opposite side, carry the ends around the head and cross them; incline the head laterally, and secure the ends of the triangle to the shoulder cravat.

Used to approximate wounds at the side of the neck.

Handkerchief Bandages of the Trunk.

Axillo-cervical cravat. Place the body of the cravat in the axilla, carry the ends over the shoulder, across each other, and around the neck.

Used to retain dressings in the axilla.

Bis-axillary cravat (simple). Place the body in the axilla, cross the ends over the shoulder and carry one across the chest, the other across the back, to the axilla of the opposite side, where they are tied or pinned.

Used as the preceding bandage.

Bis-axillary cravat (compound). Place the body of one cravat in the axilla, carry its ends over the shoulder and tie (axillo-acromial cravat). Place the body of another cravat in the opposite axilla, and carry the ends obliquely across the chest and back to the first cravat, tying them together when one end has passed through the loop of the first cravat.

Used to retain dressings in both axillas.

Bis-axillo-scapulary cravat (simple). Place the body to the front of the shoulder, with the lower end one-third longer than the upper. Carry the upper end over the shoulder, the lower end under the axilla, obliquely across the back to the opposite shoulder, around it, and back to the short end, to which it is tied. This forms a posterior figure-of-eight, and is used as a temporary dressing for fractured clavicle.

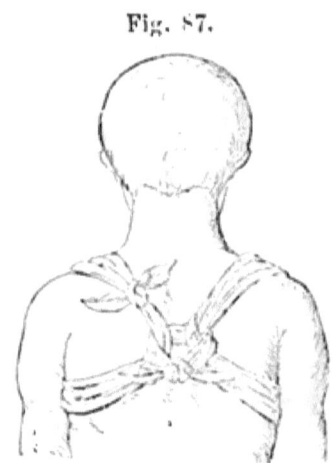

Fig. 87.

Bis-axillo-scapulary cravat (compound).

Bis-axillo-scapulary cravat (compound). Loop one cravat loosely about the shoulder, and tie. Place the body of the other cravat in front of the opposite shoulder, carry the ends back, one over the shoulder, the other through the axilla. Tie in a single loose knot, carry one end through the loop of the first cravat, and tie in a double knot.

Used to draw the shoulders forcibly back, as in fracture of the clavicle.

Dorso-bis-axillary triangle (compound). Breakfast shawl. Carry a cravat around the chest and tie in front (dorso-sternal). Place the base of a triangle, apex down, on the back of the neck, carry each end over the corresponding shoulder, and tie to the dorso-sternal cravat in front. The apex is fastened around the body of the cravat behind.

Used to retain dressings to the shoulder or back.

Mammary triangle. Place the base of the triangle under the breast, and its apex over the shoulder of the same side. Carry one end across the opposite side of the neck, the other under the axilla of the affected side. Tie at the back, and secure the apex beneath the knot.

Used to support the breast, to make pressure, to retain dressings.

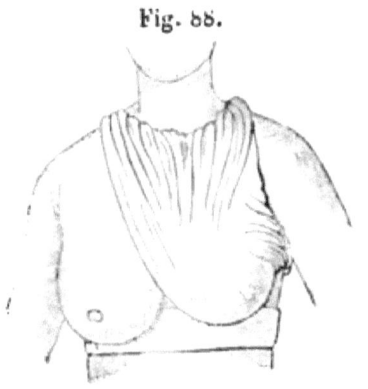

Fig. 88.

Mammary triangle.

Scroto-lumbar. Tie a cravat about the waist. Place the base of a triangle beneath the scrotum, carry the two ends up and secure them to the cravat. Finally secure the apex by carrying it under the cravat, folding it in front, and pinning.

Used as a suspensory of the scrotum.

Abdomino-inguinal (simple). For this bandage one long cravat may be made by tying two together. Place the body of the cravat back of the thigh in such a manner that one end may be two-thirds longer than the other. Bring the ends to the front, cross over the groin, and carry them around opposite sides of the body, knotting or pinning in front.

Fig. 89.

Gluteal triangle.

Used as the spica of the groin, to retain dressings on bubos, or make pressure upon them.

Abdomino-inguinal (compound). Place the centre of the cravat (three, knotted or sewed together) over lumbar vertebræ, carry the two ends forward on each side just below the iliac crests, obliquely downward and inward over the front of the groins, backward between the thighs, outward around each thigh to the front; cross over the pubes and pin to the body of the cravat.

Gluteal triangle (compound). Tie a cravat about the waist. Place the base of a triangle obliquely at the gluteal fold, and tie the ends around the thigh. Carry the apex up and under cravat, fold it over, and pin.

Used to retain dressings to the gluteal region.

Handkerchief Bandages of the Extremities.

Palmar triangle. Place the base of the triangle on either the palmar or dorsal surface of the wrist, fold the apex over the hand and back to the wrist, carry the ends around the wrist and apex and tie, fold the apex back, and pin to the body of the bandage.

Triangular cap of the shoulder. 1. Place the base on the shoulder, apex hanging down over the arm; carry the ends under the axilla, across each other, around the arm, taking in the apex, and tie. Fold the apex upward, and pin to the body of the bandage.

2. Place the base of the bandage on the upper part of the arm, with the apex covering the shoulder; carry the ends around the arm, across each other in the axilla, and up around the shoulder, taking in the apex. Fold the apex down and pin. Used to retain dressings to the upper part of the arm or shoulder.

Triangular cap of a stump. Place the base under the stump, carry the apex over its end. Secure the apex by carrying the ends around the limb, and pinning or knotting. Fold the apex up, and pin to the body of the bandage.

Cervico-brachial triangle. Sling of the arm. Place the base of a triangle at the wrist of the flexed forearm, carry the ends over the shoulders, around the back of the neck, and tie. Draw the apex back beyond the elbow, fold it posteriorly, and pin it in this position. If the triangle is not long enough, a cravat may be tied loosely around the neck, and the ends of the triangle knotted in this.

Fig. 90.

Cervico-brachial triangle.

Metatarso-malleolar cravat. Place the body obliquely across the back of the foot, carry one end around the foot, the other around the ankle, and tie in front, over the back of the foot.

Malleolo-phalangeal triangle. Place the base in the hollow of the foot. Fold the apex around the toes and in front of the ankle-joint. Carry the ends around the foot, cross on the dorsum, and continue around the malleoli; then back to the dorsum, securing here, or continuing to the side and pinning.

Cervico-tibial triangle. Carry a cravat from the top of the shoulder of the sound side to the axilla of the injured side, around the body to the point of starting, and tie. Flex the leg and place the base of a triangle on the tibia just above the ankle. Carry the ends up and tie through the cravat. Bring the apex around the knee, and pin to the body of the handkerchief. Used to support the leg when it is fractured, and the patient is required to walk.

Figure-of-eight of the knee. Place the body of the cravat just above the patella, carry the ends back, cross in the popliteal space, bring them forward just below the patella, and tie. Used to approximate the fragments of a fractured patella.

Tarso-patellar cravat. Place one cravat as a figure-of-eight of the knee, loop another cravat around the foot, just anterior to the ankle; catch the body of the third cravat through this loop, and carry its ends under both the lower and upper segments of the figure-of-eight, and secure by pinning. Used to approximate the fragments of a broken patella.

Tibial cravat. Place the body obliquely across the calf, carry the ends around the leg, one below the patella, the other above the malleoli. Used to retain dressings.

Barton's cravat. Place the body of the cravat around the point of the heel, with the end corresponding to the outer side of the foot one-third longer than the other. Hold the inner end (short) parallel with the foot, while the long end is carried across the instep, turned once around the inner end, back under the sole of the foot, and looped around itself as it crosses obliquely over the instep. The two ends are knotted, drawn upon, and the cravat so arranged that traction exerts equal pressure upon dorsum and heel. Used to make extension for fractured femur.

INDEX.

ABSCESS, acute, 27
  bone, 171
  Brodie's, 171
  chronic, 29
  diploë, 78
  follicular, 212
  mammary, 253
  mediastinal, 134
  periosteal, 169
  periurethral, 213
  residual, 30
  tubercular, 29
Amputation, 282
  Carden's, 286
  Chopart's, 284
  Dupuytren's, 289
  Gritti's, 286
  Hey's, 284
  in coxalgia, 165
  in fracture, 110
  in gangrene, 43
  in gunshot wounds, 71
  Larrey's, 289
  Lisfranc's, 284
  Pirogoff's, 284
  Sédillot's, 285
  Syme's, 285
  Teale's, 283
Anæsthetics, 258
Anchylosis, 168
  in coxalgia, 165
  in fracture, 121
Aneurism, anastomotic, 244
  arterio-venous, 74
  cirsoid, 244
  classification, 245
  traumatic, 74
  varicose, 74
Aneurismal varix, 74
Angioma, 244
Antiseptic treatment, 44, 66
Anus, artificial, 187
  diseases of, 198
  fissure, 202

Anus, fistula, 202
  malformation, 198
  pruritus, 205
  ulceration, 205
Arthrectomy, 278
Arthritis, 161
  gelatinous, 162
  rheumatoid, 167
  strumous, 162
    of hip-joint, 163
    of knee-joint, 166

BALANITIS, 212
  Balano-posthitis, 212
Bandages, handkerchief, 31
    Barton's, 308
  roller, 290
    Barton's, 299
    Désault's, 294
    Gibson's, 300
    Velpeau's, 293
Barton's cravat, 308
  fracture, 122
  head bandage, 299
Bed-sore, 41
Bites, 72
Bladder, atony, 229
  bar at neck, 226
  exstrophy, 227
  inflammation, 228
  paralysis, 229
  rupture, 227
  tumors, 227
Bone, diseases, 169
  syphilis, 173
  tubercle, 173
Brodie's abscess, 171
Bronchotomy, 250
Bronchus, foreign body, 250
Bubo, d'emblée, 211
  gonorrhœal, 213
  primary, 211
  syphilitic, 206

Bunion, 256
Burns, 102
Bursa, dropsy, 256
Bursitis, 256

CALCULI, vesical, 233
   Callus, 109
Canal, femoral, 192
   inguinal, 189
Cancrum oris, 40
Carbuncle, 42
Caries, 172
Catheter, Mercier, 231
   olive-pointed, 220
   prostatic, 231
   railroad, 220
Cellulitis, 52
Chancre, 206
Chancroid, 210
Chilblain, 249
Chloroform, 259
Chordee, 214
Cicatrization, 31
Circumclusion, 63
Clap, 211
Cleft palate, 255
Club-foot, 254
Cock's perineal section, 222
Cold, effects of, 249
Colles's law, 210
Compression, cerebral, 84
Concussion, cerebral, 83
   of lung, 94
Contusion, abdominal, 96
   cerebral, 83
   of cranium, 78
   of scalp, 77
Counter-irritation, 23
Cowperitis, 213
Coxalgia, 163
   diagnosis, 166
Cupping, 22
Cystitis, 228
Czerny's suture, 90

DELIRIUM tremens, 46
   Diffused aneurism, 74, 245
Dilatation of stricture, 220
Discharge, urethral, 215
Dislocation, *see* Luxation
Dissecting aneurism, 245
   wound, 71

Double inclined plane, 130
Dressing, Lister's, 67
Dupuytren's splint, 133

EMPHYSEMA, 94
   Encephalitis, 86
Enterocele, 180
Entero-epiplocele, 180
Epididymitis, 213
Epiplocele, 180
Epispadia, 217
Erysipelas, 50
Ether, 258
Excision, 278
   ankle-joint, 281
   elbow-joint, 279
   hip-joint, 279
   in coxalgia, 165
   knee-joint, 280
   shoulder-joint, 278
   wrist-joint, 279
Extension apparatus, 129
Extravasation, intracranial, 81
   of urine, 223

FÆCES, impaction of, 204
   False joint, 111
   passage, 219
Fever, hectic, 50
   inflammatory, 48
   pyæmic, 49
   septicæmic, 48
   traumatic, 47
Fissure, anal, 202
   of Rolando, 88
Fistula, anal, 202
   fæcal, 187
   salivary, 90
Forcipressure, 62
Foreign body in brain, 87
   in bronchus, 250
   in larynx, 250
   in œsophagus, 252
Fractures, 105
   anæsthetics in, 112
   Barton's, 122
   clavicle, 114
   coccyx, 126
   Colles's, 122
   compound, 108
   delayed union in, 110
   delirium tremens in, 136

INDEX. 311

Fractures, diagnosis, 107
    femur, 126
    fibula, 132
    humerus, 117
    hyoid bone, 114
    inferior maxilla, 113
    larynx, 114
    metacarpus, 125
    nasal bone, 112
    non-union in, 110
    patella, 131
    pelvis, 125
    phalanges, 125
    Pott's, 132
    radius, 122
    ribs, 134
    sacrum, 126
    scapula, 116
    skull, 78
    Smith's, 122
    sternum, 134
    superior maxilla, 113
    T, 117
    tarsus, 134
    tibia, 132
    treatment, 107
    ulna, 121
    ununited, 110
    vertebræ, 135
    vicious union, 111
Fracture-box, 133
Frost-bite, 249
Furuncle, 41

GANGLION, 256
    Gangrene, 38
Germ theory, 44
Glanders, 55
Gleet, 215
Gonorrhœa, acute, 211
    chronic, 215
    in women, 216
Granulations, 31
Gumma, 208

HÆMATOCELE, 239
    Hæmaturia, 220
Hæmophilia, 177
Hæmothorax, 93
Hare-lip, 255
Hemorrhage, 57
    arrest of, 58

Hemorrhage, bladder, 230
    kidney, 230
    urethra, 230
Hemorrhoids, 199
Hernia, 179
    cerebri, 87
    classification, 180
    congenital, 188, 191
    crural, 192
    encysted, 188, 191
    femoral, 192
    incarcerated, 182
    infantile, 188, 191
    inflamed, 182
    inguinal, 188
    irreducible, 181
    Littré's, 184
    of lung, 94
    reducible, 180
    strangulated, 183
    umbilical, 194
Herniotomy, 186
Hutchinson's teeth, 209
Hydrarthrosis, 161
Hydrocele, 238
Hydrophobia, 54
Hypertrophy of prostate, 225
Hypospadia, 217

IMPACTED fæces, 204
    Imperforate anus, 198
Impermeable stricture, 222
Incarcerated hernia, 182
Incontinence, urinary, 232
Inflammation, 17
    intracranial, 86
Ingrowing toe-nail, 257
Internal strangulation, 197
Intestinal obstruction, 196
Intussusception, 197

KYPHOSIS, 178

LAPAROTOMY, 100, 198
    Laryngotomy, 251
Larynx, foreign body, 250
Leeching, 23
Lembert's suture, 99
Ligament, coraco-humeral, 143
    Y, 150

INDEX.

Ligamentous union, 135
Ligations, 261
    anterior tibial, 276
    axillary, 268
    brachial, 269
    common carotid, 263
    dorsalis pedis, 277
    external carotid, 264
    external iliac, 272
    facial, 265
    femoral, 272
    internal mammary, 268
    lingual, 265
    occipital, 266
    palmar arches, 271
    popliteal, 274
    posterior tibial, 275
    radial, 270
    subclavian, 267
    temporal, 266
    ulnar, 271
Litholapaxy, 235
Litholysis, 235
Lithotomy, 235
Lithotrity, 235
Localization, cerebral, 87
Loose bodies in joints, 167
Lordosis, 178
Luxations, 137
    astragalus, 156
    carpus, 148
    classification, 137
    clavicle, 141
    complications, 139
    femur, 150
    humerus, 143
    jaw, 140
    metacarpus, 149
    old, 139
    patella, 155
    phalanges, 149
    radius, 148
    ribs, 141
    scapula, 143
    semilunar cartilages, 155
    tarsus, 156
    tibia, 154
    treatment, 139
    ulna, 147

MALIGNANT pustule, 55
    Meningitis, 86
Micro-organisms, 14

Mortification, 38
Mucous patch, 207

NÆVUS, capillary, 244
    venous, 244
Necrosis, 172
Nodes, periosteal, 169
Noma pudendi, 41

ŒSOPHAGUS, foreign body, 252
    stricture, 252
Onychia, 257
Ophthalmia, 213
Orchitis, 240
Osteitis, 170
    deformans, 170
    rarefying, 170
Osteomalacia, 174
Osteomyelitis, 170
Osteoporosis, 170

PAGET'S disease, 253
    Paraphimosis, 212
Paronychia, 257
Passage of catheter, 219
Perineal section, 222
Peritonitis, 97
Periostitis, 169
    osteoplastic, 169
Pernio, 249
Phimosis, 212
Phlebitis, 242
Piles, 199
Plaster jacket, 176
Plastic lymph, 18
Pneumothorax, 94
Pneumotomy, 95
Pott's disease, 174
    puffy tumor, 78
Poupart's ligament, 190
Prolapsus of lung, 94
    recti, 201
Prostatitis, 224
Pruritus ani, 205
Pupil in brain injury, 85
Pus, 19
Pyæmia, 49

RACHITIS, 176
    Rectum, diseases of, 198

Rectum, polyp, 204
    prolapse, 204
    stricture, 203
    ulceration, 203
    villous tumor, 205
Resection, 278
Retention of urine, 230
Retroclusion, 62
Rheumatism, gonorrhœal, 213
Rickets, 176
Ring, abdominal, 189, 190
    femoral, 193
Rupture (see Hernia), 179
    of viscera, 96

SALIVATION, 26
    Saphenous opening, 193
Sarcocele, 240
Sayre's fracture-dressing, 115
Scalds, 102
Scalp, layers, 75
    wounds, 77
Scoliosis, 178
Scrofula, 177
Septicæmia, 48
Shock, 45
    ether in, 260
Sinus, 30
Skin grafts, 57
Spine, curvature, 177
Splints, coxalgia, 165
    Bond's, 124
    Dupuytren's, 133
Sprain, 158
    fracture, 158
    of back, 159
Staphyloplasty, 255
Staphylorraphy, 255
Stimulants, 25
Stings, 72
Stone in bladder, 233
Strapping chest, 135
Stricture, urethra, 217
Struma, 177
Sutures, 65
Synovitis, 160
    gonorrhœal, 213
Syphilis, 206

TALIPES, 254
    Tapping abdomen, 100
        bladder, 232

Tapping pericardium, 95
    pleura, 95
Taxis, 184
Tenosynovitis, 256
Tetanus, 52
Thrombosis, 242
Torsion, 61
Torsoclusion, 62
Trachea, foreign body in, 250
Tracheotomy, 251
Transfusion, 59
Trephining, 89
Triangles of neck, 262
Trophic changes, 75
Tubercle, 173
Tumors, breast, 253

ULCERATION, 31
    Ulcers, 32
Uranoplasty, 255
Urethra, 211
    deformities, 217
    rupture, 223
    stricture, 217
Urethrotome, 221
Urethrotomy, 221

VARICOCELE, 240
    Varicose aneurism, 74
        veins, 243
Varix, 243
    aneurismal, 74
    arterial, 244
Veins, diseases of, 242
    varicose, 243
Venereal disease, 206
Vesication, 24
Volvulus, 197

WALLERIAN degeneration, 74
    White swelling, 162
        hip-joint, 163
        knee-joint, 166
Wounds, 44
    abdomen, 95
    arteries, 73
    chest, 92
    classification, 68
    contused, 69
    dissecting, 72
    face, 90

Wounds, gunshot, 70
    incised, 69
    joints, 159
    lacerated, 69
    neck, 91
    nerves, 75
    œsophagus, 92

Wounds, poisoned, 71
    punctured, 69
    scalp, 77
    trachea, 92
    veins, 75
    Y ligament, 151

www.ingramcontent.com/pod-product-compliance
Lightning Source LLC
Chambersburg PA
CBHW030747250426
43672CB00028B/1255